ELECTRICITY IN AGRICULTURE AND HORTICULTURE

Prof. Selim LEMSTRÖM

With illustrations

First Edition:
"THE ELECTRICIAN" PRINTING &
PUBLISHING COMPANY, LTD.,
Salisbury Court, Fleet Street, Lokdon
1904

THERMOMETER SCALE.

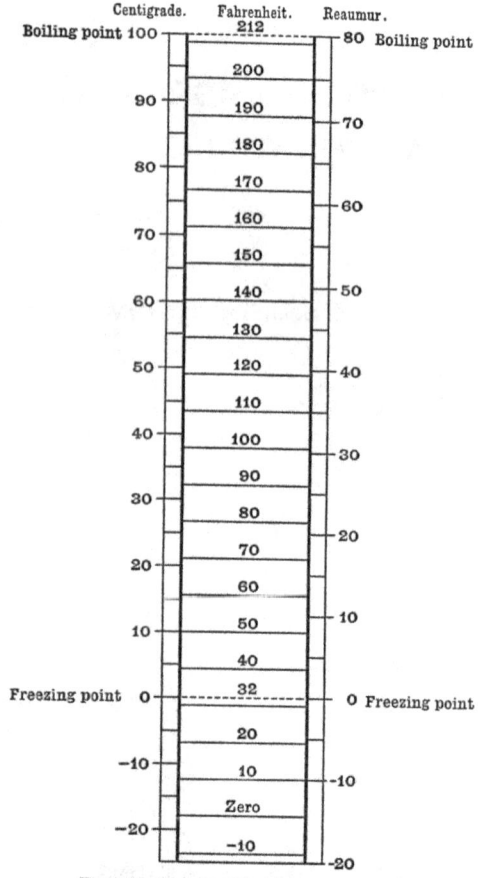

Formulæ for converting from one system to another.

$$F = \frac{9C}{5} + 32.$$
$$C = \frac{5(F-32)}{9}.$$
$$R = \frac{4(F-32)}{9}.$$

PREFACE

'The Electrician' Printing and Publishing Company has had the kindness to undertake the publishing of *Electricity in Agriculture and Horticulture*, which, I venture to hope, will be received favourably by the English-speaking public who read it. It contains the results of a long study and numerous experiments which had their origin in the Polar regions, were continued in more southern latitudes, and have led not only to an increase of the harvest of every kind of plant which has come under treatment, but also to a favourable change of their chemical compounds, *eg.*, an increase of the digestible nitrogenous matter in seeds, of the sugar in sugar-beets, proved by chemical analyses, and of the sweetness in berries, &c. The earlier ripening of some berries and fruits is also a proved result. There have been many difficulties to overcome, including the English language; but here the Publisher, Mr. Geo Tucker, has kindly given me assistance by reading the manuscript for the press, for which I am very thankful. The like gratitude I would express to Mr. R. B. Greig, for much good counsel and many corrections.

A great difficulty has been to estimate the cost of the apparatus and the quantities of some of the articles necessary for the electrical installation to enable me to apply the electric air current to growing crops, especially the cost of wood-posts, of labour, of inspection, &c, which are of varying value in different countries; but as these quantities form in the total cost only a small part, this does not affect the calculations in a serious way.

THE AUTHOR.

PUBLISHER'S NOTE

The Publisher is desirous of placing Prof. Lemström's interesting work before the public, with the view of hastening the time when the utilisation of the electric current in agricultural and horticultural work will become a recognised development of the science and industry of up-to-date cultivation of the land for market crops. That this development will be advanced by Prof. Lemström's valuable experiments there can, the Publisher considers, be no question.

As the greater part of the Author's scientific determinations regarding temperatures are made in Centigrade degrees, and for other measurements in the metric system, the Publisher, in accord with the Author's wishes, and desirous of facilitating the reading of this book, has printed tables containing a Figure showing the relations between the different thermometers (page 4), and a Table showing the metric measures (with their signs) in English measures (on last pages). The Table is attached in such a manner that it can be readily consulted by the reader during his reading of the book.

INTRODUCTORY

§ 1. It is well known that the question which is the subject of this book has been a favourite field of investigation for a century past. As the subject is connected with no less than three sciences—viz. physics, botany and agricultural physics—it is in itself not particularly attractive. The causes which induced me to begin an investigation of this matter were manifold, and I venture to hope that a short exposition of them will not be without interest.

During several voyages in the Polar regions (1868 to Spitzbergen and the north of Norway, and 1871, 1882, 1884 to Finnish Lapland), I had occasion to observe with my own eyes a peculiarity in the vegetable kingdom which also has attracted the attention of other explorers. When the plants in these regions have resisted the frequently destructive night frosts they show a degree of development which greatly surpasses that of plants in more southern regions, where the climatic conditions are more advantageous. This rich development appears principally in the fresh and clear colours of the flowers, in their strong perfume, in the rapid development of the leaves on the trees, and their scent, but particularly in the rich harvest which different seeds—such as rye, oats and barley—will produce when, as before stated, they are not destroyed by the frosts. From a bushel of rye sown they will often produce 40 bushels, and from barley 20 bushels, and so forth. It is the same with grass. These results are attained although the people cultivate their soil very imperfectly, using only ploughs and harrows of wood.

Vegetables require, for a rich development, a fertile soil and a sufficient quantity of heat, light and humidity; but one of these principal conditions, that of heat, is to be found only in a slight degree in the Polar Regions. To be convinced of this it is only necessary to glance at a map and follow the yearly isotherms, and especially the summer isotherms. The isotherm corresponding to a mean temperature of -7° C to -8° C *crosses* Spitzbergen, and that of 0° C goes through the north of Finnish Lapland. The mean temperature for the month of July there is 10° C, and at Spitzbergen 5° C.[1]

The cause of the astonishing development of vegetation under such conditions has hitherto been held to lie in the long day, which lasts from two to three months in these regions during their short summer. It is further sometimes supposed that this circumstance will provide the crops with more light and heat, the principal factors of vegetation.

---

1. *See* Thermometer Table on page 4.

This is, however, not true, for calculation shows that the quantity, notwithstanding this long day, is less than it is, for instance, in 60 deg. of latitude. In consequence of the low height of the sun over the horizon, the rays meet the earth in such an oblique direction that their illuminating and heating power is greatly lessened, and this power is also lessened by the great absorption in the atmosphere. Since the quantities of light and heat received are relatively smaller than in more southern regions, the cause of the rich development of the vegetables cannot lie in the long day, and must be sought elsewhere.

For several reasons I was induced to search for the cause in electrical currents, the effect of which appears in the Polar light, and which are going from the atmosphere to the earth, or vice versa. The existence of these currents has been proved beyond all doubt through the experiments of the Finnish International Polar Expedition of 1882-84.

§ 2. My reasons will be shown in the following order:

(a) The physiology of plants gives a satisfactory explanation of the functions which most of their organs have to perform, and good reasons for their existence and their varying forms. This is, however, not the case with the needle-like shape of the leaves in fir trees, and the beard on the ears of most cereals. Since nothing exists without a purpose in all the infinite number of objects in Nature, then the needle-shaped leaves and the beard must have their determined ends. In fact, they are very well fitted to be the means through which the electricity goes from the atmosphere into the earth, or vice versa; that is to say, they act in the same fashion as metallic points. To pretend that they really serve as a means of transmitting electricity only because their form shows them capable of it would be to go too far. The presence of electricity in the air around them shows that they are, in fact, in a position to perform this function of transmission. Through experiments made during the above-named expedition, it was, at least by analogy, proved that they really serve that end, for it was not only shown that such electrical currents exist, but these currents were even measured by means of a specially constructed apparatus provided with metallic points.

We are thus induced to concede that an electrical current is going on through the needle-formed leaves of the pine, and the beard on the ears of cereals, not to particularise other plants. Though this is the case, it is not, however, shown that this electrical current exercises any influence whatever in the process of vegetation, or, in other words, has any definite effect on the plant life itself. That must be proved by experiments, and a description of them, with an exposition of their results, is the principal object of this volume.

(b) In studying and observing sections of fir trees from different latitudes (60 deg. to 67 deg.) we have found a peculiarity in their yearly growth in thickness or year rings. They show in general a great difference in growth during different ages, but when due regard is paid to this fact, so that these differences may be eliminated, there remains another difference, which clearly must depend on the more or less advantageous nature of the atmospheric conditions of the year.

These latter variations, taking into account principally the thickness of the year rings, show a periodicity in full agreement with the periods of the sun spots and the auroras—namely, a period of from 10 to 11 years. By a comparison between sections of large pine trees from the Polar regions of 67 deg. of latitude and from more southern latitudes (about 6o deg.), it will be seen that this periodic variation is much more pronounced the more nearly we approach the Polar regions. The circumstance that this periodicity agrees with that of the auroras will conduce to an investigation of a possible connection between the electrical current producing auroras and the year rings of these trees. As this peculiarity is more strongly developed in those regions of the earth where the electrical currents in question are of greater frequency and higher intensity, it indicates a connection between cause and effect which demands an investigation.[1*]

(c) In a short Paper entitled, *On the Periodic Variations in some Meteorological Phenomena, their Connection with the Changes of the Solar Surface and their Probable Influence on the Vegetation*,[2] I have suggested, with a high degree of probability, that the harvest results in Finland show a periodicity which agrees with periodical variations in the sun spots and in the number of Polar lights. The greater the yearly number of sun spots and auroras, the more abundant is the harvest of seeds, roots, and grass.[3]

In the Paper in which I treated these phenomena I have found the explanation of this periodicity in the different state of the heat radiating from the sun, either without spots or supplied with them. In the former case, the greater part of the heat rays are light and of shorter wave length; in the latter case, the greater number are dark and of greater wave length. The rays must pass through the atmosphere, and the darker heat rays will be absorbed during this passage rather than the lighter ones, and hence there will be more heat stored up in the

---

1. This investigation is not yet published.

2. Finsk tidfkrift, 1878, in Swedish.

3. These harvest results also give evidence of the longer cycle period of about 58 years.

atmosphere in the case when spots appear on the sun. This heat, generally contained in watery vapours, is transported through the air currents or winds into higher latitudes and will there exercise its beneficent effect. Though I continue to regard this explanation as highly probable, I must introduce some modification, in order to ascribe a large influence to the Polar light, or rather its cause, the electric current from the atmosphere.

In consequence of this modification of my ideas, I must give it an important place in order to explain the periodicity of the harvests.

§ 3. Everyone who has given attention to the question of the causes of the electrical conditions of the atmosphere knows that several theories on this matter have been brought forward. The most widely known suppositions are the following:

(a) Unipolar induction, caused by the earth, as a magnet, rotating around its axis, produces a force component, which carries the positive electricity up in the atmosphere.

(b) Evaporation, which, according to the opinion of several physicists, produces electricity, the vapours being positive.

(c) Unipolar induction working together with the evaporation in making every particle, great or little, carried from the earth, positively electric.

(d) The vegetation process on the earth; the friction

of the small solid or liquid particles suspended in the atmosphere with the air, or between each other, the air's friction against the surface of the earth.

(e) The direct effect of the sun rays on the different layers of the atmosphere.

A discussion of these very different views cannot be entered upon here. I will, therefore, only say that the majority of the physicists, as it seems to me, are at present inclined to give a certain precedence to the opinion which searches for the cause of the atmospheric electricity in the evaporation which is going on all over the earth (perhaps in connection with the unipolar induction). For my part, I have no doubt that this conception is right—viz. that the watery vapours act as transmitters of the atmospheric electricity and carry it to the upper layers of the air, because it seems to me that this theory is very well founded and in full harmony with other closely allied phenomena.

According to this opinion, therefore, the evaporation must be subjected to the same periodicity as the sun spots, and also the electrical phenomena in the atmosphere. My opinion upon the manner in which Nature is fulfilling all this process is, shortly, as follows[5]:

The quantities of electricity which are carried up by the water vapours to the higher parts of the

---

5. *See* also *L'Aurore boréale*, by S. Lemström. Paris: Gauthier Villars, 1886, p. 131 and following.

atmosphere reach, there, an air stratum with a low pressure. As this rarefied air is provided with a relatively high conducting power, it will form with the surface of the earth a nearly spherical condenser. The rarefied air space lies, principally in consequence of the lower temperature, nearer to the earth surface around the poles. Through this circumstance, a greater quantity of electricity will accumulate in these regions, and discharge in auroral displays or in a current from the atmosphere.

It has already been announced that the Finnish International Polar Expedition produced, by experiments made on mountain tops, auroral beams and light phenomena of the same nature as the auroras, and hence contributed to the confirmation of the opinion that auroras are caused by electric currents in the atmosphere. The EMF (electromotive force), or the force which produces this current, was measured. It was absolutely very little, but always existing, though with varying intensity. It is only in exceptional cases that it will produce light phenomena—viz. when the conditions of the atmosphere are favourable, and the current is of high intensity. The most important result here is the proof of the existence of such a current, which is always passing either downwards or upwards.

The experiments which have been carried out in Finnish Lapland, do not, strictly speaking, apply to other regions of the earth. Considering, however, that the laws of atmospheric electricity are nearly the same for all latitudes, the existence of such a current in the whole atmosphere seems not to be in doubt.

This electric atmospheric current has hitherto been very little investigated; it is only the atmospheric electricity which has been measured. The method has been to measure the potential in a point of the air (or the electrical tension of it). These results cannot be used for conclusions regarding electrical current from the atmosphere, and, in consequence, not for the laws it follows. These observations of atmospheric electricity can, however, contribute to some general propositions. When it has been found, for instance, that the air near the earth's surface is positive electrically (and seldom negative), and that the potential increases with the height, a general conclusion can be drawn that an electric current, an equalising of the difference of potential, is going on between the whole atmosphere and the earth.

As has been said above, this current produces luminous phenomena or Polar light only when in a state of high intensity. When these manifestations appear, principally in the Polar Regions, it must be admitted that the current there possesses the highest intensity. The effect of it must, therefore, be more remarkable there than elsewhere.

Once convinced not only of the necessity of finding a cause for the above-mentioned phenomena of the vegetable kingdom in the Polar regions, but also of the existence of the electric current from the atmosphere, I was strongly induced to connect the two phenomena, and to regard the electric current as the cause of the peculiar phenomena in the vegetation, and all that has been said about these peculiarities will find its explanation in this: — (1) The rich harvests and their periodicity; (2) the periodical increase in the year rings of the fir woods; and (3) the needle-formed leaves and the beards of the ears will, from this point of view, only be the means by which the electricity is conducted from the atmosphere to the earth. As the intensity of the current is highest in the Polar Regions, its effect there must also be the greatest.

It must not be overlooked that electricity has hitherto been regarded as of little or no importance in the complex life of a plant, and a great effect from its application was not anticipated. From a further consideration of the subject it will, however, be clear that this conception must be changed, and that electricity must be numbered among the principal factors in plant life.

THE EXPERIMENTS

§ 4. Since the year 1746, when the Scotchman Maimbray made his experiments on the influence of electricity on two myrtles, a great number of researches have been performed to examine this influence. The most striking feature of these experiments is that they are *always contradictory*. Hardly has one explorer obtained favourable results before another presents himself with directly opposite ones. A closer scrutiny shows that a favourable result is obtained every time the experiments are executed with artificial electricity—namely, with a machine. Exception shows only one case in which negative electricity was employed. The experiments have never been executed on a large scale, because the method was not convenient. In the case when the method was intended to spread the electricity over a larger field its application seemed to present much difficulty.

The cause of the contradictory results has been searched for under conditions of only occasional occurrence—for instance, difference in soil, different illumination, &c. Without denying that such circumstances can exercise a very great influence, my more recent experience has shown that these contradictions can be explained without reference to chance. Consideration of all the evidence I have obtained shows so decidedly that electricity exercised a favourable influence on the growing plants, that I found a new examination of the question necessary.

My first experiments, the aim of which was to test the usefulness of the method, were carried out in the physical laboratory of the University of Helsingfors. Omitting a number of experiments, the results of which only encouraged me to continue further, I will enumerate the first definite experiments made from the beginning of May to 24 June 1885:

(1) Before a window looking to the south, three small compartments of cardboard were constructed, and on a table within were placed two flower pots in each compartment. In these pots were sown barley, wheat and rye—four grains of each sort. The grains of each kind were of the same weight and appearance. Over the pots was hung a net of wires provided with points, separated from each other and insulated. The soil in the pots was connected with the earth by tinfoil in such manner that, when we call the compartments I., II., III., the electric current from the positive pole of a Holtz electric influence machine was going in:

    I. From the points in the air to the soil.

    II. From the soil in the pot to the points.

    III. Without wires.

The negative pole of the machine was conducted to earth.

The pots were watered with water of the same temperature and of the same quantity in each pot. After certain intervals, the plants were measured in height, and the leaves in both breadth and length.

A week passed. We could already see a remarkable difference. The plants in the compartments I and II developed much more strongly and more rapidly than in III. The electrical machine was kept working five hours daily. The experiments were continued to 24 June, when the increase of vegetation in I and II under the influence of the electric current was estimated to be 40% beyond the result in III. As the soils in the six pots were exactly alike, the cause of the greater development *must* be sought in the electric current.

The difference between I and II was barely perceptible, though the current in I was + and in II −.[6] No difference between + E and − E is shown in these experiments. (*See* also next page on this point).

(2) In the summer of 1885, the first experiments were performed on a field. This was a small barley field on the farm of Mr. W. Lemström, in the parish of Vichtis, in the south of Finland.

Though the external circumstances were unfavourable, dryness being prevalent,[7] the results

---

6. + stands for positive current, − for negative current.

7. That dryness would be unfavourable was not known at this time, but this will be seen later on.

were very satisfactory, the part of the field under the electric current giving an increase of about 35.1%.

(3) In the summer of 1886, the experiments were executed on a garden field belonging to the Garden Society in Helsingfors. The plants came under treatment in such a way that half a garden bed of about 7 metres (23 feet) in length was under the current, the other half serving as "control." The following are the results in percentages:

—	Per cent Increase	—	Per cent Inc. or Dec.
White beets	+ 107.2	Garden strawberries in a greenhouse	**
Potatoes	+ 76.2		
Red beets	+ 65.29	Carrots	- 5.12
Radish	+ 59.1	*Turnip-cabbage	- 5.23
Parsnips	+ 54.45	Cabbage	- 43.58
Onions	+ 42.11	*Turnip-cabbage	+ 1.8
Celery	+ 36.90	Turnip	+ 2.58

* Kohl Rabi   ** See below

A singular circumstance was that we could not, during the summer, see any difference between the experimental and the control field, and we were about to believe that no difference at all would appear in the harvest, but, from the table above, it seems that the difference was very great.

The most remarkable fact in this table is that some of the plants seem favoured by the electricity while others derived no benefit, and were apparently injured by it. Among these last mentioned, cabbage is prominent. It is, however, not necessary to discuss this inequality here, as it was later discovered that it had its cause in the want of water.

The strawberries planted in pots were stored in a greenhouse and arranged in three compartments, separated from each other by walls of cardboard. In each compartment were two pots, in the first + electricity was given and − electricity in the second, the third being without electricity. The berries ripened in the first batch in 28 days, in the second in 33 days, and in the third in 54 days. The effect of the current was, therefore, to shorten in an astonishing manner the time of ripening. Here we have a marked difference between the + E and the − E, the former having a greater stimulating effect.

In the same summer (1886), a first experiment was performed on the estate of Brödtorp in the parish of Pojo, south of Finland. I had the good fortune to interest the owner, the Baron E. Hisinger, in these experiments, and I am under a great obligation to him for his assistance and guidance with the utmost courtesy and kindness. The experiments, executed on a wheat field of exceptionally good development, first began on 20 June, when the wheat was just flowering. When the crop was sifted in first and second qualities the result was:

—	**Experimental Field**	**Control Field**
First quality	198.1kg	126.1kg
Second quality	1,028.1kg	1,166.4kg
Total	1,226.2kg	1,292.5kg

(both fields were of 0.5 hectare)

As the experiments began as late as the time of flowering of the wheat, an increase of the total crop could not be expected, only an improvement in quality. And that is what happened: we can see that the first quality is increased on the experimental field by 57%.

(4) As the experiments hitherto performed had, as shown above, given very encouraging results, we were induced to make an extensive series of experiments at Brödtorp in the year 1887, to which Baron Hisinger consented and gave, as before, his friendly assistance.

We give below the results obtained, and refer the reader to the extensive treatment of this subject in *Expériences sur l'influence de l'électricité sur les Végétaux*[8]:

---

8. *Commentationes variae in memoriam actorum CCL annorum*. Edidit Universitas Helsingforsiensis, 1890; and in the *Electrical Review* (London), 4 and 25 November and 2 and 16 December, 1898.

		Average increase per cent.
Rye ... $\begin{cases} 3\cdot7\% \\ 15\cdot7\% \end{cases}$	The experiments came about so late that the ears began to be formed...	10·0
Wheat $\begin{cases} 51\cdot4\% \\ 39\cdot5\% \end{cases}$		45·1
Oats		54·0
Barley		85·0
,,		45·0
Potatoes		24·3
Red beets		31·7
Raspberries		95·1

As in the experiments which took place in the field of the Garden Society at Helsingfors, a number of plants showed even here a decided decadence. The most prominent were:

Plants	Decrease (%)
Carrots	- 47.9
Peas	- 47.1
Swedish turnips	- 35.5
Tobacco	- 27.4
Flax	- 6.7
Cabbage	- 4.5

The varieties were partly the same as in the field of the Garden Society, the new ones being flax, tobacco and peas.

(5) The results obtained by the experiments made up to 1887 were very encouraging to a continuation of the studies of the phenomenon; but still many problems were yet to be solved. Among these were the following:

Do we obtain the same results as in Finland if the experiments should be performed after the same method in another part of the globe, or, in other terms, would the effect be the same independent of the latitude?

Considering the great importance of the whole question, any publication of the results without having given an answer to that question would, perhaps, have been fatal. The first and most important point of all was to raise the necessary funds. That great difficulty was overcome and, by Prof. E. Mascart, we were introduced to Baron A. Thenard, who gave us permission to perform the experiments on his estate at Castle Laferte, in Burgundy, and also his support where it was needed.

Of the exposition hitherto made, it seems that the greatest difficulty with experiments in the open field is to get a homogeneous soil on both the experimental and control fields. But that is not all. The fields must also be uniformly lighted by the sun. The shadow of a tree can be of great consequence if it falls unequally on both fields, and that circumstance will be more effective in the case of a summer with, in general, bad weather. The summer of 1888 was exceptionally bad, and fatal to the experiments on a number of plants, because the field on which they were carried out was not suitable in many cases. It has, therefore, been necessary to exclude a number of doubtful results; but for further details we must refer to a more complete publication, *Expériences sur l'influence de l'électricité sur les végétaux*.[9] Here, we give only the results which were certain and of consequence for our purpose.

The plants which were objects of suitable experiments were:

In the garden: Raspberries, peas, onions, carrots and cabbage.

In the field: Wheat, oats, maize, red beets and white beets.

The results gave an augmentation in the following proportions:

Plants	Increase (%)
Wheat	about 21.2
Oats	about 18.6
Maize	about 2.6
Raspberries	about 42.8
Red beets	about 16.4
Peas	about 75.0
Beans	about 36.4

---

9. *See* note on page 25.

Among the plants treated were strawberries, which we must consider separately, because the experiments on them were most instructive. On the experimental field were 315 plants, and on the control field 392 plants. The experiments began on 18 April 1888, and were continued day and night. On the plants were, at the beginning, only leaves, but on 29 April appeared the first flowers on the experimental field, and eight days later on the control field. During this interval of 19 days there were four days of rain, and, as the machines are not effective under conditions of rain and moisture, the time of active action was reduced to 15 days. The development of the flowers on the experimental field was, therefore, twice as rapid as on the control field, and the plants of the former field possessed an appearance of much greater activity than the latter. This state of things continued to 18 May, but then, followed eight days with a burning sun and exceptional warmth, even for Burgundy. From this time, the plants on the experimental field began to languish, and the berries gathered between 7 and 21 June gave:

– Experimental field: 8,065 kg, or 157 g. from one plant

– Control field: 7,245 kg, or 185 g. from one plant or 15.1% less from the experimental field.

A difference in quality was also noticeable, the berries from the control field being fresher and more fragrant, those from the experimental field sweeter. It was evident that the electricity had damaged the plants. In order to estimate rightly the damage, the number of fruits was counted to be an average of 16 per plant on the experimental field, but only eight on the control field; the fecundity had been twice as great on the experimental field. The cause of this unexpected change in the development must be sought in the electric current, and the consequence is that the quantity of this agent must be lessened in times of burning sun. But it will be seen that an equal watering of both fields should have altered the result in favour of the experimental field.

We must then conclude that electricity is not so harmless that it may be given in an unlimited quantity; the external circumstances must be carefully considered. The same damaging through the burning sun happened with maize, which, in the beginning of May, showed an evident greater activity on the experimental field than on the control field, but was found afterwards to decline.

In all the series of experiments, the results showed that some of the plants were not favoured by electricity. Among these were carrots, cabbage, rooted cabbage, tobacco and peas.

During the experiments in Burgundy, besides the areas upon the field, there were fields of experiment and control both for carrots and peas in the garden. These four small fields being of the same size were equally watered; care was taken that every field should receive the same quantity of water. The result was that the carrots gave an increase of 125%, and the peas (after careful estimation) 75%.[10]

The results so far showed, therefore, that if electricity is applied according to the above-mentioned method, favourable results were obtained equal to an increase of 40 to 80%. Besides, it has been shown that electricity administered at the time of strong sunshine was damaging the development of the plants. One other experience gained from these experiments is the following:

The more fertile the soil and the more vigorous the vegetation, the more stimulating will the effect of the electric current prove; a number of results show this. We will compare two experiments on potatoes and red beets, the one made on the field of the Garden Society, the other on the field at Brödtorp. The former field possessed all the properties of garden soil, the latter was a little better than an ordinary farm land:

Plants	Increase	
	Garden soil	Field
Potatoes	76.2%	24.3%
Red beets	65.3%	31.7%

---

10. The peas were, shortly before ripening, purloined by pigeons.

Electricity will, in a high degree, accelerate the ripening of fruits, berries and roots, and probably develop more sugar in them. The former effect is well shown by experiments on strawberries and raspberries, the latter will be the object of further investigation.[11]

In the greenhouse, the strawberries under electrical treatment ripened in an average time of 30 days, while 24 more days were necessary for the berries without the benefit of the electric current, or a total of 54 days. It was the same with raspberries on the open field in Brödtorp, though the difference of time there was only of 17 days. The development of sugar has been stated, in an analysis made in France on white beets, to give an increase of about 18% of sugar,[12] to which the remarkable augmentation of the sweetness of the strawberries in the same place is additional evidence.

Comparing the experiments made in southern Finland and in Burgundy, we find the results in general to be the same, but the increase seems to be less important in Burgundy. We must, however, remember that the weather during the summer of 1888 in Burgundy was exceptionally bad, and, considering the law of greater development in the

---

11. See the analysis of sugar beets in the experiments of the summer of 1902 and 1903 on page 84.

12. See results of the analysis on page 84.

more vigorous vegetation, we cannot assert that the fact of less increase per cent. in southern regions is sufficiently proved.

§ 5. *Experiments continued.* — After a long interval, the experiments were resumed in 1898. Some of the questions still remained unanswered. Amongst these was the question: What time of the day is best for applying electricity, and are the effects proportional to the time during which it is applied?

During the interval between the experiments, a new type of influence machine had been constructed, with cylinders instead of plates, rotating in the opposite direction. This type of apparatus, which has since been modified again,[13] came in use in the summer of 1898, when the experiments were carried out at Villa Kammio, near Helsingfors, where Dr. W. Lybeck had kindly given a piece of land in his garden and a room for the machine. In these experiments, the soil for the experimental field, as well as for the control field, was divided into two parts, whereby its properties were equalised. The machine was in motion from 5 am to 9 am, and from 4 pm to 8 pm, or about eight hours daily.

The experiments began on 17 June, and by the middle of July tobacco plants on the experimental field showed so evident a superiority that

---

13. *See* Fig 5. page 40.

photographs of both fields were taken, one of the experimental field and one of the control field. Two of these photographs are represented in Figs. 1 and 2 on the opposite page.

The machine had been at work for 164 hours when the photographs were taken. The difference between the experimental and control fields was estimated at 40%. The experiments were continued until 24 August, and the harvest gave:

– from the experimental field: 16 plants weighing 18 kg;

– from the control field: 16 plants weighing 13 kg, or nearly 39%, better result on the experimental field.

The fields were situated on gently sloping ground and were divided into two sections, an *upper* and a *lower* one, of which the upper was watered with 50% more water than the lower. The results were the following:

– Carrots: an increase of 8.7%
– White beets: an increase of 11.2%
– Beans: an increase of 11.1%.

Of the carrots, the upper and more watered field showed the above increase; of the beets, on the contrary, the lower or less watered. The remaining fields showed nearly the same result. The short time (only eight hours daily) during which the machine was in motion exercised a considerable influence on the results.

FIG. 1.—Experimental Field.

FIG. 2.—Control Field.

We must here call attention to a singular circumstance concerning these experiments. It seems that the tobacco plants had by middle July, or during the time I could personally survey the experiments, already given, approximately, the surplus that was attained in the final result. Therefore, it appears as if the electricity had no increased effect over the interval of more than a month during which the experiments were continued. This strange circumstance gives rise to the supposition that a defect, not seen by the person in charge, belongs to these experiments. I suppose that the machine had been running uncharged during the greater part of this time without this fact being observed; I was myself on a journey.

§ 6. The following year, in the summer of 1899, the experiments were again removed to Brödtorp, where Baron E. Hisinger, with his usual benevolence and never-failing interest, gave us some land and caused the necessary work for the culture of the plants to be continued. As stated above, the greatest difficulty in such experiments, when a comparison between the harvests from two fields has to be made, lies in obtaining homogeneous soil—that is, a soil which fully possesses the same qualities of fertility. As this is dependent not only on the elements of which it is composed, but also on the depth of cultivation and humidity of the soil, choosing the land is very difficult. In order to overcome these difficulties, at least to a certain degree, the following method was introduced. It will be best illustrated through Fig. 3 below:

C_I	E_I		C_{II}		E_{II}		C_{III}
Cu	Cu	Cu	Cu	Tr	Tr	Tr	Tr
Go_3	Sb_3	Go_3	Sb_3	Ca_3	Ca_3	Po	Po
Sb_r	Go_r	Sb_r	Go_2	Po	Po	Ca_2	Ca_2
Go_r	Sb_2	Go_2	Sb_2	Ca_r	Ca_r	Po	Po
R_3	R_3	Sb_3	Sb_3	Tr	Be_3	Pe_3	Pe_3
Sb_r	Sb_r	R_2	R_2	Pe_r	Pe_r	Be_2	B_r
R_r	R_r	Sb_2	Sb_2	Be_r	Be_r	Pe_2	Pe_2
W_3	W_3	St_3	St_3	O_2	O_3	B_3	B_3
St_r	St_r	W_2	W_2	B_r	B_r	O_2	O_3
W_r	W_r	St_2	St_2	O_r	O_r	B_2	B_2

FIG. 3.

The plants taken under experiment were:

Cereals: Wheat = $W_1\ W_2\ W_3$; Rye = $R_1\ R_2\ R_3$; Oats = $O_1\ O_2\ O_3$; Barley = $B_1\ B_2\ B_3$

Shell Vegetables: Peas = $Pe_1\ Pe_2\ Pe_3$; Beans = $Be_1\ Be_2\ Be_3$

Roots: Sugar Beets = $Sb_1\ Sb_2\ Sb_3$; Carrots = $Ca_1\ Ca_2\ Ca_3$; Potatoes = P_o

Fruits: Gooseberries = $Go_1\ Go_2\ Go_3$; Strawberries: $St_1\ St_2\ St_3$; Currants: C_u

Clover = T_r and Cabbage[14] = $Sb_1\ Sb_2\ Sb_3$.

---

14. Cabbage was in the summer of 1900 sown on the lower squares marked Sb1 Sb2 Sb3.

The tobacco was grown in fields which, in August 1899, were sown with rye, and the raspberries were in fields where the wheat was sown. The former had little, the latter no success at all. The columns with superscription C_I, C_{II}, C_{III} signify control fields, those with E_I and E_{II} experimental fields. As shown by Fig. 3, every sort of plant had three experimental and three control fields.[15] Each field was 5 square metres (16.5 feet), and the different control and experimental fields were separated by a space 5 metres broad sown with oats. These fields were reaped green.

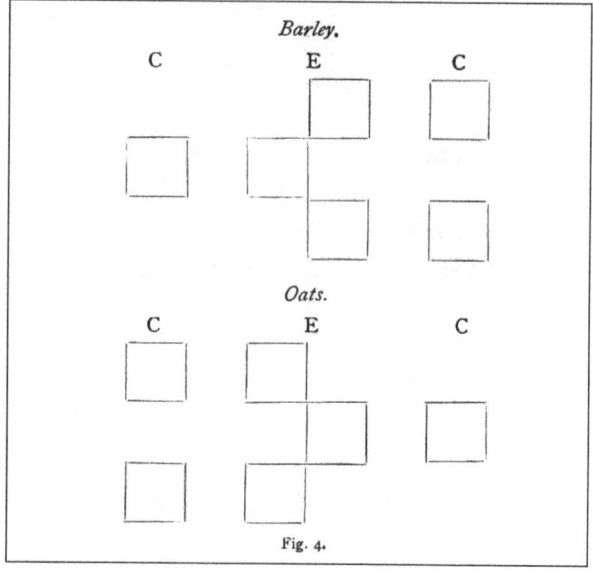

Fig. 4.

---

15. Except currants, clover and beans, where some fields were missing.

Among the plants which were under experiment during the first year, results could not be expected from currants, gooseberries, strawberries and clover, or from wheat and rye, which were sown on 21 August. For the most sensitive plants, such as sugar beets, carrots, beans and cabbage, the soil during 1899 was not sufficiently cultivated, and so a result from them could not be expected, particularly when we remember that the greatness of the percentage of increase depends on the vigour of the vegetation. There remained this year, therefore, only barley, oats, peas and potatoes.

One might think that all inequalities in fertility would be eliminated when the experimental squares and the corresponding control squares are situated in the manner shown in Fig. 4, where the squares under C are control fields and under E experimental fields; but that was not the case, for the soil represented in Fig. 3 under the last column (C_{III}) showed such a superiority over the soil in the columns marked C_{II} and E_{II}, that the given results would, without observing this fact, be erroneous. Therefore, to obtain a reliable result, it will be necessary to graduate the soil according to its fertility.

If to all the squares under E_{II} the value 1 is given, the squares under C_{III} must be put = 1.4, and the squares under C_{II} = 0.8. The harvest from the squares under C_{III} must be divided by 1.4 and under

C_{II} by 0.8. In this way, we shall attain, not the reality, but a minimum of increase.

By this graduation we had a good proportion of oats growth in the fields which lay between C_{II} and E_{II} and between E_{II} and C_{III}; the growth in the latter was at least twice the former.

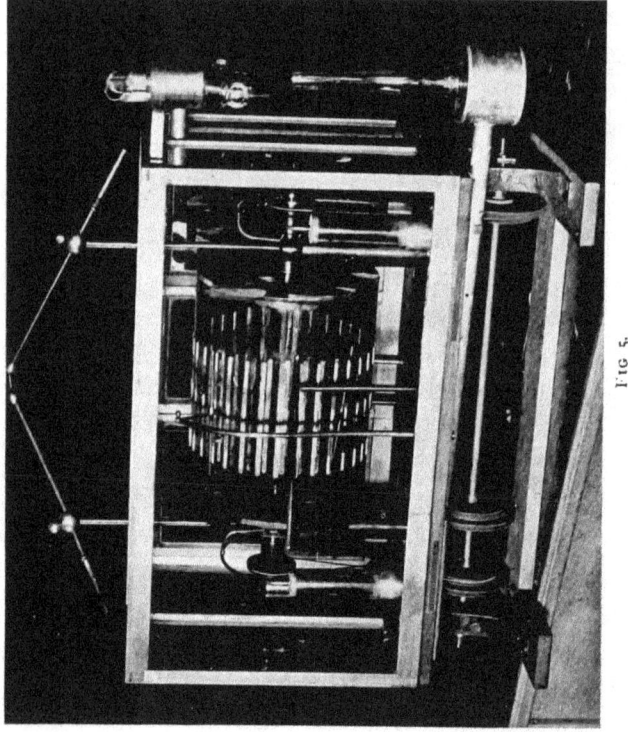

Fig 5.

The new influence machine was used on these experiments (see Fig. 5), with cylinders of about 30 cm. in diameter and 40 cm. in length. It was kept in motion about eight hours daily, from 7 am to 11 am and from 4 pm to 8 pm. This rule could not, however, be regularly followed, especially in humid weather. In case of rain, the machine ceased to work. On the other hand, the machine was at work during a longer period in cloudy days, when it was not to be feared that electricity with sunshine would exercise a damaging effect. In such a way, the machine worked:

– from 15 to 30 June, an average of 6.5 hours daily,
– from 1 to 15 July, an average of 6.5 hours daily,
– from 16 to 31 July, an average of 6.3 hours daily,
– from 1 to 15 August, an average of 7.7 hours daily,
– from 16 to 31 August, an average of 6.3 hours daily,
– from 1 to 15 September, an average of 6.7 hours daily,
– from 16 to 24 September, an average of 6.2 hours daily.

The fields were sown on 14 and 15 June, 1899. Watering, as equally as possible, was done on 28, 29 and 30 June, 4, 5, 7, 14, 18 and 19 July, and on 2 August. Rainy days were during 4 July; 11 August; 11 September. The insulated wire net was laid out round the fields E_I and E_{II} in such a way that a galvanised iron wire of 1.5 mm. in diameter was drawn on posts round the field, and on this wire were extended cross-wires at a distance of 1.25 metres from each other. The wires were fixed on the posts with sheltered insulators of ebonite.

RESULTS

Oats

Experimental fields		Control fields	
Number	Weight	Number	Weight
1	34.9kg	1	30.0kg
2	38.3kg	2	-
3	39.5kg	3	27.9kg

Average increase: 28.7%.

Barley

Experimental fields		Control fields	
Number	Weight	Number	Weight
1	42.1kg	1	37.7kg
2	36.1kg	2	31.9kg
3	42.1kg	3	29.0kg

Average increase: 23.0%.

Peas

Experimental Fields		Control Fields	
Number	Weight	Number	Weight
1	41.2kg	1	42.7kg
2	35.3kg	2	39.3kg
3	35.5kg	3	39.1kg

Average decrease: 7.5%.

Cabbage

Field number	Experimental fields			Control fields		
	Number of plants	Weight	Weight of a plant	Number of plants	Weight	Weight of a plant
1	48	53.16kg	1.167kg	43	38.70kg	0.899kg
2	50	57.37kg	1.448kg	48	73.94kg	1.541kg
3	48	48.88kg	1.018kg	43	69.05kg	1.606kg

Average decrease: 19.4%.

Sugar beets

Field Number	Experimental fields		Control fields	
	Leaves	Roots	Leaves	Roots
1	17.21kg	11.05kg	26.35kg	13.18kg
2	26.70kg	18.28kg	17.85kg	9.78kg
3	17.64kg	11.05kg	29.33kg	13.60kg

As the number of plants was equally distributed over all the fields, we can here compare the average of the weight of the roots. This gives an increase of 10.4% for the experimental field; but as the growth in the roots depends upon, besides other circumstances, the greater or less number of leaves which the plant is spreading in the air, it is preferable to compare the squares where the weight of the leaves was nearly alike—that is, the experimental fields 1 and 3 and the control field 2, and likewise experimental field 2 and control fields 1 and 3. The former gives an increase of 13.0%, and the latter 38.8%.

Carrots

Only two squares could be compared with each other, in consequence of the unequal development in the other fields.

—	**Experimental fields**	**Control fields**
Number of plants	469	502
Leaves	6.38kg	4.86kg
Roots	18.91kg	14.72kg
Weight of a root	0.0403kg	0.0293kg

Showing an increase of 37.5%.

Potatoes

Field number	Experimental fields				Control fields			
	Number of plants	Leaves	Roots	Average harvest of a plant	Number of plants	Leaves	Roots	Average harvest of a plant
1	175	27.63kg	53.13kg	0.27kg	175	21.04kg	36.13kg	0.18kg
2	187	26.14kg	48.24kg		190	19.13kg	34.91kg (48.87kg)	
3	197	23.38kg	49.30kg		193	19.13kg	30.36kg (42.50kg)	

50% of increase[16].

The remaining plants gave no results, partly because this was their first year of growth, partly from other causes. The beans, for instance, were destroyed by night frosts.

When, in considering these experiments, we take into account the imperfect cultivation of the soil and the late sowing, they show, in general, the same results as during the previous years' experiments.

The decreasing tendency of cabbage and peas has been observed before, and shows only that the watering has been too little, which was the case at the beginning of the experiment during the month of June.

The experiments made in the year 1899 might, in general, be regarded as a preparation for the following year, 1900.

As the growth was under the average, the consequence must be a low increase per cent., according to the law stated above. This has really

---

16. The numbers in parentheses are those actually received.

been the case. Remembering that the electricity was applied only during 6.5 hours daily, we must regard the results as satisfactory.

§7. The experiments were continued during the year 1900, and were chiefly undertaken for the purpose of finding out the effect that electricity exercises *during the night*.

As the wheat and rye were this year observed with special attention from the sowing onwards, we will devote to them a separate description. From 1899 there were in the fields gooseberries, currants, strawberries and clover, which, like the other squares, were manured. The fields were also thoroughly cultivated, beginning in the autumn of 1899.

– The sowing of barley and oats was done on 30 May.

– The sowing of potatoes was done on 5 June.

– The sowing of carrots was done on 6 June.

– The sowing of beans and sugar beets was done on 7-8 June.

The germination in the barley and oats squares began at the same time on both the experimental and the control fields. The germination of sugar beets, carrots and potatoes occurred at the same time, and the flowering also was contemporaneous on both fields, so that regarding these, no inequality was observed.

The influence machine was at work all the time it could be charged from 2 to 18 June, and likewise, from 6 to 13 September. During the rest of the time, it was at work with small changes, from 7 pm to 7 am.

Period	Average number of hours daily	Length of spark in mm		Number of days	Weather	Electricity
		Day	Night			
2-18 June	17.0 day and night	1.5 to 2.2	0.5 to 1.9	8	—	—
19-30 June	10.5 night	0.5 to 1.5	0.2 to 0.8	—	Sky often clear	Alternating + pole and - pole every day
1-15 July	8.6 night	1.5 seldom	0.3 to 0.5	4	Clear and half clear	
16-31 July	10.7 night	1.5 to 0.8 seldom	0.1 to 0.8	1	Cloudy and clear	
1-15 August	7.2 night	Very short		1	Misty	
16-31 August	7.6 night	—	0.1 to 0.5	3	Clear and half clear	
1-13 September	12.5 night and day	1.5	0.1 to 0.3	5	Cloudy and half clear	

The spark-length between the discharging balls was measured* at the time of setting in motion, when it was stopped, and sometimes, while it was going on. In general this length is very variable, as will be seen from the above. The spark-length is the measure of the potential to which the wire net can be charged, and depends on external meteorological conditions, the most significant of which is the humidity of the air. The faster the electricity is streaming out from the wire net the lower is the potential to which the wire net can be loaded by means of a machine of given strength. The electricity meets in such a case with smaller resistance in the air-layer between the wire net and the plants. This resistance is, in general, greater in the day, the air being drier, than at night, when the air is more humid. The electrical air-current which goes from the wire net through the plant is, therefore, different during the day than during the night. The plants show in their life cycle essential inequalities during the day and the night. It is for this reason that the electrical treatment was in 1900 restricted to the night hours.

Barley

Field number	Experimental fields		Control fields	
	Straw	Seed	Straw	Seed*
1	7.0kg	1.950kg	4.25kg	1.686kg (1.350kg)
2	8.0kg	1.800kg	7.00kg	1.926kg (2.200kg)
3	6.5kg	2.850kg	8.00kg	1.608kg (2.250kg)

Increase: 26.4%.

---

* The numbers actually received are put in parentheses.

Oats

The oat fields showed this year such great inequality that no sure results were to be expected from them. They were situated a little higher on the squares where the carrots were planted in 1899. This inequality was due to an old ditch, which some ten years previously had been filled. This ditch went through two of the experimental squares, and lowered their fertility in such a remarkable way that comparing them with the other squares with a prospect of sure results was out of the question. My opinion is, however, as I had an opportunity of following the whole development during the summer, that the electricity exercised so great an influence that the effect of it was about the same as on the barley above.

Peas

Field number	Experimental fields		Control fields	
	Straw	Seed	Straw	Seed
1	7.5kg	0.810kg	4.0kg	0.506kg (0.405kg)
2	4.5kg	1.003kg	5.0kg	0.716kg (1.003kg)
3	5.0kg	1.205kg	6.0kg	0.716kg (1.003kg)

Increase: 55.7%.

Potatoes

Field number	Experimental fields		Control fields	
	Number of plants	Weight of reaped potatoes	Number of plants	Weight of reaped potatoes
1	150	19.30kg	128	16.50kg
2	121	20.50kg	112	(37.70kg)*
3	127	17.40kg	—	—

If the control field 2 is excluded, the increase will be seen to be 17.0%, a number very little different to that obtained previously. So the result must be regarded as very uncertain.

Carrots

Field number	Experimental fields		Control fields	
	Leaves	Roots	Leaves	Roots*
1	4.950kg	13.500kg	2.900kg	8.125kg (6.500kg)
2	8.000kg	20.800kg	4.500kg	9.428kg (13.200kg)
3	6.800kg	17.700kg	5.200kg	9.429kg (13.200kg)

Increase: 92.7%.

---

* *See* note p. 48, as well as for following charts.

Sugar beets

Field number	Experimental fields		Control fields	
	Leaves	Beets	Leaves	Beets
1	14.0kg	19.70kg	10.15kg	15.30kg
2	14.3kg	21.10kg	16.50kg	18.20kg
3	11.6kg	23.20kg	11.50kg	11.50kg

Increase: 42.2%.

Clover

The number of experimental fields was two and of control fields three. One of the experimental fields suffered from the same cause as mentioned above concerning the oats squares, and must, therefore, be left out of consideration. If the second control field is excluded, so that only the two control fields situated on both sides of the experimental field come into consideration, the results will be:

Field number	Experimental field		Control fields	
	Hay	Seed	Hay	Seed
1	10.0kg	2.20kg	8.7kg	2.20kg
3	—	—	8.0kg	1.68kg

The average gives an increase of 19.8% of hay, and 13.4% of seed.

We must, however, regard this result as uncertain, because the control field 1 showed from the beginning a high degree of development, and had to be excluded from comparison. Control field 3 was again situated in the column of squares the fertility of which was too great. When we take that into consideration, and use the same minimum number of reduction as above, or 1.4, the increase will be:

75.1% of hay and 83.3% of seed.

The uncertainty is, however, great in this experiment, and the percentage of increase can hardly be determined.

Beans

Field number	Harvest	Experimental fields		Control fields	
		Weight in grammes	Total weight	Weight in grammes	Total weight*
1	29 August	855g	2,005g	390g	1,625g (1,300g)
	9 Sept.	1,150g		910g	
2	29 August	1,250g	2,420g	2,850g	(4,070g)
	9 Sept.	1,170g		1,220g	
3	29 August	1,985g	3,265g	1,725g	2,232g (3,125g)
	9 Sept.	1,280g		1,400g	

If the control field 2 is excluded, as the fertility is too great for comparison, and if No. 1 is divided by 0.8 and No. 3 by 1.4, the increase will be 36.6%.

At the same time, it seems that the first average harvest from the experimental field will surpass that from the control field if the latter is treated in the way recommended, which shows an earlier development on the former.

Currants

On the squares planted with currants, the number of bushes which had begun to grow in the experimental field and in the control field was very different, so that, regarding the quantity, no comparable results could be obtained.

Gooseberries

Of gooseberries, on the contrary, the following harvest was obtained:

Field number	Experimental fields		Control fields	
	Harvest	Weight in grammes	Harvest	Weight in grammes
1	23 August	137g	23 August	62g
2	23 August	175g	23 August	112g
3	23 August	75g	23 August	169g

Increase: 12.8%.

That the effect of electricity has been great is, however, seen from fields 1 and 2, where the experimental fields have a great overweight. That the control field 3 has such a great overweight over the corresponding experimental field results from the greater development of a couple of bushes. Excluding this number, the increase will be 79.3%.

Strawberries

Harvest	Experimental fields			Control fields		
	Field 1	Field 2	Field 3	Field 1	Field 2	Field 3
27 July	—	86g	25g	—	41g	—
27 July	—	25g	—	—	—	—
1 August	—	42g	6g	—	26g	—
3 August	10g	23g	12g	3g	8g	10g
8 August	11g	13g	12g	5g	11g	19g
11 August	13g	11g	20g	—	16g	11g
14-27 August	—	22g	14g	3g	18g	12g
Total	34g	222g	89g	11g	120g	52g

Though the harvest cannot be regarded as good, either from the experimental field or the control field, the increase on the experimental field is of such superiority that the effect of the electricity is evident. The percentage of increase is, on the average, from all three fields, 88.7%. The earlier development is also noticeable.

§ 8. On 21 August, 1899, the rye and wheat were sown in rows. The sowing began at 6 am, and was finished at 7 pm, in the following order:

– Control Fields: W_1 W_3 R_1 R_3
– Experimental Fields: R_3 R_1 W_3 W_1 W_2 R_2
– Control Fields: R_2 W_2

On 28 August, the first rye-germs were observed; unfavourable weather delayed the germination.

On 29 August, the rye-germs in the experimental fields were much longer and more numerous than on the control fields.

On 30 August, the wheat had come up in all the experimental fields, but only a few germs were observed in the control fields.

On 31 August, the influence machine was stopped.

On 1 September, the germs were carefully observed in all the fields. In the experimental fields they were remarkably better and more numerous than in the control fields.

The difference in germination time was 12 to 24 hours.

In a square of 40 cm^2., the number of germs were counted and their lengths measured as follows:

Wheat

Field number	Experimental fields		Control fields	
	Number of germs	Average length	Number of germs	Average length
1	48	10.9cm	55	10.3cm
2	59		49	
3	69		50	
Total	176		154	

Rye

Field number	Experimental fields		Control fields	
	Number of germs	Average length	Number of germs	Average length
1	71	11.5cm	47	10.8cm
2	65		54	
3	58		58	
Total	194		159	

The growth on the experimental fields appeared tighter and more developed.

During the summer of 1900 the experimental fields proved to be superior to the control fields both for wheat and rye. On 10 July in that same year, both rye and wheat lay down so that the wire net came to stand higher than it ought to have done. The harvest was made on 16 August, and gave the following results in grammes:

Wheat

Field number	Experimental fields			
	Straw and seed together	Seed		
		First Quality	Second Quality	Total
1	24,800g	6,000g	1,620g	7,620g
2	24,500g	5,700g	1,900g	7,600g
3	27,200g	3,770g	1,500g	5,270g

Field number	Control fields			
	Straw and seed together	Seed		
		First Quality	Second Quality	Total
1	20,550g	4,770g	1,300g	6,070g
2	22,250g	4,370g	2,700g	7,070g
3	20,800g	4,600g	3,150g	7,750g

As it seems that the experimental field 3, which gave the greatest quantity of straw and seed together (27,200 g. against 24,500 g. and 24,800 g. in both 2 and 1), has given the least quantity of seed, a mistake has clearly been made. The most probable error is that a clerical mistake has been made in writing 3,770 g. instead of 5,770 g.

The experimental fields 1 and 2 give an average of 5,800 first quality seed. The control fields 1 and 2 give an average of 4,570 first quality seed or an increase of 28% of first quality seed. The combined increase is 16.8%.

Rye

Field number	Experimental fields			
	Straw and Seed together	Seed		
		First Quality	Second Quality	Total
1	19,900g	3,460g	2,860g	6,320g
2	20,600g	2,380g	4,350g	6,730g
3	18,750g	3,670g	2,710g	6,380g

Field number	Control fields			
	Straw and Seed together	Seed		
		First Quality	Second Quality	Total
1	20,500g	2,900g	3,260g	6,160g
2	15,800g	2,250g	2,560g	4,810g
3	16,650g	2,050g	3,110g	5,160g

The experimental fields give an average of 3,170 first quality seed. The control fields give an average of 2,400 first quality seed or an increase of 32.1% of first quality seed. The combined increase is 28.4%.

§9. Comparing the results for both the years 1899 and 1900, they appear to be alike for barley and oats, or an increase of 25 to 30%; for peas the decrease of 7% has grown to an increase of 55.7%; and for carrots from 37.5 to 92.7%. In general, the percentage of increase was greater during the summer of 1900. During the summer of 1899 the machine was at work, on an average, for 6.5 hours *daily*; during the summer of 1900, principally during *the night*, for 10 hours. The longer time during which the machine was at work in the latter summer will suffice to explain the higher increase, so we cannot conclude from these results whether day or night is more favourable for the electrical treatment.

When all the results received are taken into consideration, the increase appears greater if the application of electricity is continued by day and night. So we seem entitled to suppose that there is a certain dependence between the percentage of increase and the period during which electrical treatment takes place; but any declaration that they are proportional to each other cannot be made

from the experiments hitherto carried out.

As above mentioned, the electric current which goes from the wire net through the plants to the earth is different during the day and the night: during the day, high potential and great resistance; during the night, low potential and little resistance. It must be understood that we here take normal conditions for granted—that is to say, clear days and clear nights. During moist weather the potential is low, and in case of rain, the machine can continue to work only when a nearly constant spark is going on in a gap (1 mm.) of the circuits to the earth. This was not applied in the above-mentioned experiments, but the machine was stopped.

It might now happen that the changes in the potential and the resistance were such that the intensity of the current was constant, or nearly so; but upon this point, our knowledge is at present very inadequate. Most probably the intensity in question is very variable. Hence it seems that this relation between the increase per cent., and the period during which the electrical air-current exercises its effects is also a variable quantity, and that the effect of the electric air-current is different under different exterior conditions. This must be taken into consideration in judging the amount of the increase percentage.

As mentioned above, the greatest difficulty in determining this percentage of increase lies in the want of uniformity in the soil. An augmentation of the number of squares might perhaps lead to an equalising of the difference in the average fertility of the soil; but this will lead to another difficulty—namely, that which consists in making out the many squares. Much skill and foresight are necessary to avoid confusion.

As will be seen from the experiments previously described, there were for rye and wheat 12 squares (six for each). It would have been necessary to prepare each square separately in order to attain positive results. That mistakes might happen here is easy to comprehend, and some have really happened; but they were so apparent that they could be immediately corrected. To prepare the whole soil beforehand would obviously have been too costly.

A relatively satisfactory result could certainly be obtained in the following way: Two squares of 3 m. (10.8 sq. ft.) are prepared with the soil from the field in which the electrical treatment is to be applied. They must lie at least 5 m. from this field and from each other. The soil on both the squares must be well mixed and put out in equal quantities in every square. One is put under the same electric treatment as the field. If care is taken that the sunshine and the rain will reach both the squares in the same proportions, comparable results can be obtained. The increase per cent. for the whole field can then be calculated.

The experiments in the summer of 1903, particularly at Åtvidaberg, Sweden, showed it to be necessary to improve the method of control, as the constitution of the soil will have a marked influence, especially in rainy summers. Besides the above-described method, it should have a wholly artificial addition of well-mixed earth put into two boxes,

1 m². in surface and 1/2 m. in depth, the sowing being as equal as possible in number and at the same depth. One of the boxes should be placed under the electric air-current, and the other, l0m apart from the field under the same current.

The growth in these boxes will show the effect of this current, and also provide the means to judge if the fields have been of equal fertility.

In connection with these experiments, it is to me an agreeable duty to express my best thanks to Baron E. Hisinger for the warm interest he has taken in the work and for the real support he has given to it.

HOW DOES ELECTRICITY EXERCISE ITS INFLUENCE ON THE PLANTS?

§10. During the summer of 1898, some experiments at the Physical Laboratory of the University of Helsingfors were performed regarding the state of liquids in capillary tubes under the influence of the electrical air-current. These experiments were continued in the spring of 1900 by myself, assisted by a student (R. Bengelsdorff), and in the autumn and winter of 1901 with the aid of another student (V. J. Laine). The most important of these results may here be cited.[17]

17. *See* the Author, *On the State of Liquids in Capillary Tubes under Influence of Electrical Air-Currents*, Ofversigt af Finska Wet.

If a capillary tube is lowered into a glass of water united electrically with the earth, and a fine metal point, in conducting connection with the *negative* pole of an influence machine, is placed above the capillary tube, when the machine is started water-drops will appear after some seconds in the upper part of the tube. It is assumed that the inner surface of the tube has been moistened just before, and that the positive pole of the machine is united with the earth. The water climbs up the walls of the tube and forms one or more drops, or small water cylinders. The quantity of water, which in such a way ascends in the tube, is proportional to the square of the distance of the point from the meniscus of the water pillar in the tube, and is dependent upon the length and diameter of the tube and the resistance in the circuit.

The positive pole exercises no influence, so that this effect is produced by the current from the earth through the liquid in the capillary tube to the point. When the connection with the earth is removed, all effect ceases. Hence it follows that an electric current is produced by which the liquid is drawn up in the positive direction.

The application of this phenomenon on the method used in our experiments on growing

---

Soc., Forh. Bd. XLIII., 1901. In German : Drudes Annalen. 1901, S Bd., p. 729.

plants is as follows: The electric current, produced by the influence machine when its positive pole is earthed, goes from the earth through the plants to the points of the insulated wire net and back to the negative pole. This causes an ascent of liquid or juice in the capillary tubes of the plants, and *produces in such a way an augmentation of the energy with which the circulation of the juices is going on*.

During the experiments made hitherto, the insulated wire net had been united with the positive pole of the influence machine, and had thus been made positive; in analogy with Nature itself, where the potential of the air is ordinarily positive. A thorough investigation of the influence machine resulted in detecting the following peculiarity, not before observed by me: As the machine, provided with Leyden jars at both poles and with the one pole earthed, is stopped, after having been again moved and set in movement afresh, there will occur a change of poles. This change easily escapes the attention of the observer. Such a change of pole might, therefore, have occurred during the experiment by which the wire net had been charged alternately positively and negatively.

For this reason it was very important to make a series of experiments in which strict control over the direction of the current was exercised. The experiments were executed in the same way as the experiments mentioned on pages 21-23, with, however, a few modifications.

Sixteen burnt clay pots of middle size (19 cm. in diameter at the upper side, 14 cm. in diameter at the bottom, and 19 cm. in height) were placed before the windows, four in number, facing south. The pots were placed four in a line before the window. The earth in these pots had been well mixed in a box and distributed alike in all the pots. In each of 12 pots out of the 16 was sown, on 18 May, 1901, 12 grains of wheat, the same of oats and barley—two grains in the middle, and in five symmetrically situated places also two grains. In the four remaining pots were set strawberry plants in the middle, and around them in three places were sown carrots. The pots were placed in four sections, one before each window, so as to stand in a row—the pot with the strawberries and carrots inside, then the pot with wheat, and last that with oats and barley. Every section was separated from the other by screens of white pasteboard, and care was taken that the light was the same for all.

Above the three first sections (I, II, III) was suspended by a thread of silk a wire net furnished with points 25 cm. broad and 120 cm. long, at a height of 40 cm. above the earth in the pots. Through the holes in the bottom of the pots were introduced equal-sized strips of zinc which were in conducting communication with each other, and with the earth in every section. The pots stood on plates of burnt clay, and these on plates of thick glass supported by small pieces of ebonite. In Section IV, which ought to serve as control, the pots with their plates stood on the same kind of insulating layer, but with no zinc strips beneath, nor wire net above.

After having found a new way to determine the situation of the poles at the influence machine, we could easily control this each time the machine was set in motion. When the machine is allowed to work without Leyden jars, and its poles are at a distance of 1.5 cm., a stream of light can be observed between the poles. This fine light is violet at the negative pole, ending with a bright point on the surface of the sphere; but it is white and very bright at the positive pole. The change of colour in the light-current does not take place in the middle, but is nearer to the positive pole.

The influence machine used on this occasion was of the new type, with small cylinders (15 cm. diameter and 20 cm. in length), and enclosed in a glass box, in which was placed a little plate with sulphuric acid for the drying of the air. The machine was set in motion by an electrodynamic motor, into which was led a current of from five to six Daniel elements. The negative pole of the machine was placed in connection:

– In Section I, with the zinc strips underneath the

pots, the wire net of the same section being led off to earth;

– In Section II, with the wire net above the pots, the zinc strips of this section being led off to earth;

– In Section III, with, a commutator of ebonite (lying on a plate of glass) through which a leading connection could be produced, one day with the zinc strips, and the other day with the wire net, in the same section. At the same time the commutator effected a connection so that one day the wire net, and the other the zinc strips, were united with the earth;

– In Section IV, no current.

The machine was kept in motion from 6 to 7 am until 9 to 10 am, and from 4 to 5 pm to 8 to 9 pm. In cloudy and cold weather, which occurred very seldom, the machine was kept moving a longer time, sometimes even all night.

We worked thus from:
 – 21 to 31 May: 9.0 hours daily
 – 1 to 30 June: 9.9 hours daily
 – 1 to 31 July: 5.7 hours daily
 – 1 to 13 August: 8.2 hours daily.

From the middle of June, the windows were kept open during the night in order to lessen the warmth. Watering was done every day in measured quantities. For the first two weeks, Section II, or the section in which the current went from the plants to the wire net, showed the highest development; but from this time, Section I, where the current went in the opposite direction, began to take the ascendency, and kept it until 13 August, when the experiments ceased. This refers only to the barley, oats and wheat.

On 10 July appeared the first ear of oats in Section III, on 14 July in Section II, and on 16 July in Section I.

On 23 July the brush of a barley-ear began to appear in Section I, and on 25 July in Section II. The wheat could not, however, be brought to form ears.

This proved generally that these kinds of cereal could not be brought to a normal condition of development in the prevailing high temperature, and so the experiments ceased, as I have said, on 13 August.

But the development of the strawberry plants went on normally, except in Section I, where the plant died, probably from the strong effect of the current. In Sections II and III, the commencements of the flowerings, of which the plants in Section II showed a powerful growth, were removed.

At the end of the experiments, the cereals showed, at an exact estimation, in:

– Section I: 60% more development than in Section IV.

– Section II: 45% more development than in Section IV.

– Section III: 40% more development than in Section IV.

(*See* also Fig. 6 on opposite page).

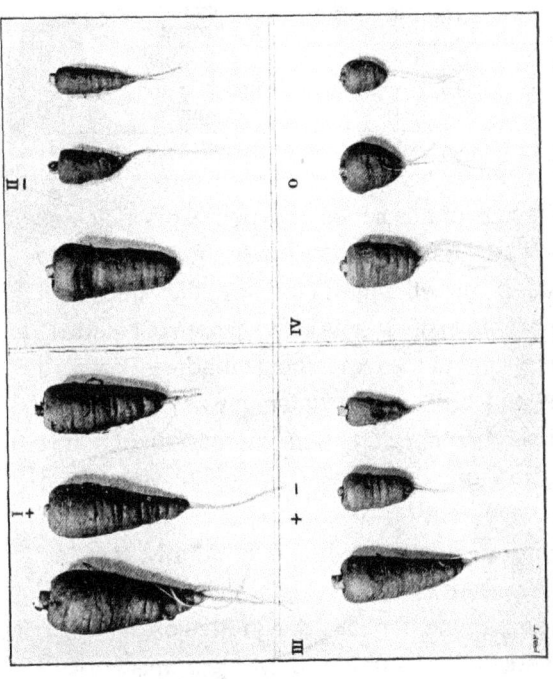

Fig. 6.—The above figure, from a photograph, shows the development of the carrots in the different compartments. The Roman figures indicate the different compartments and the signs +, −, + −, the direction of the electric current and o no current.

With regard to the strawberry plants, the same progress was observed with both the Sections II and III. The carrots showed the following results:

Section	Leaves weight in g.	Roots weight in g.	Electricity	Increase
I	30.90	104.60	Positive current	182.7%
II	15.70	49.00	Negative current	32.4%
III	16.70	42.05	Alternately positive and negative	13.2%
IV	16.65	37.00	No current	—

The results of this series of experiments indicates clearly that carrots, which have showed hitherto decreased growth from the electric treatment (except at the experiments in the garden of Laferté in Burgundy and at the experiments made at Brödtorp in 1899 and 1900, where watering was done during the experiments), were here increased; and this conclusion can be applied to other vegetables which have previously showed a decrease from the electrical treatment. It shows, further, that the positive current in Section I is giving a result which far surpasses the negative one in Section II.

Consequently, it is not only by a mechanical updrawing from the root of the water and dissolved matter within it that the electric circuit operates; it aids the development more powerfully when it goes in the opposite direction.

The question therefore arises: Is the positive current able to exercise a mechanical effect on the plants? Without doubt; for it is known that when a cut branch from a leafy tree is put into water upside down, it can suck up the water and keep living a long time; and it is the same with plants. The positive current introduces into the plants the different elements or compounds of the air, besides oxygen and nitrogen—water, carbonic acid, ozone and nitric acids, ammonia, &c.[18] A portion of these compounds is formed by the passage of the electricity from the points into the air, and the current introduces them in a fresh condition into the capillary tubes of the plant. In this way the effect of the electric air-current can be reduced to a purely mechanical one.

We do not maintain that any other effects than the mechanical may not occur; but our experience does not go so far that we can say anything with certainty on this point. The effects of the two currents can be determined in the following way: The negative current going from the earth to the plants facilitates the drawing-up of water with the dissolved matter through the roots of the plant to

---

18. Experiments executed in the laboratory of physics at the University of Helsingfors show that the electric current which is going out from the points forms the following chemical compounds, the nature of which has been determined by chemical analysis: Ozone (*in great quantities*), nitric acids, nitrous acid, ammonia (doubtful).

its capillary tubes, and thus produces a stronger circulation of the saps.

The positive current brings to the plants the different elements, and introduces them through the openings in their capillary tubes in such a way as to promote vegetation. This latter effect is, as mentioned above, very much more effective than the former.

In the above-related experiments, the effects were produced by an artificial electric air-current; but, as also mentioned above, a continual electric current is going upwards or downwards, certainly in the Polar Regions and in our latitudes, but probably everywhere; and this current must evidently exercise a greater or lesser influence on the life of vegetables in the way we have endeavoured to show.

Electricity takes a much larger part in the life of vegetables than it has been presumed to do until now. The amount or value of this influence cannot be determined without examining more exactly the electric currents of the atmosphere. The methods used at present for this purpose cannot lead to this knowledge. New methods must be introduced, and this question will be one of the most significant in the scientific programme of the future.

We cannot say that the question regarding the influence of the artificially produced electric air-currents has reached its solution; but the experiments have led to an admissible explanation, founded on facts, showing the way in which this influence is exercised. They have also effected some advance in our knowledge, which has not yet attained the necessary degree of exactness, but which possesses some importance both to science and to the practical application of our scientific knowledge. We may therefore summarise our present knowledge as follows:

(a) The real increase per cent., due to electrical treatment has not yet been exactly determined for the different vegetables which have been under experiment. But we are approaching its smallest value in fixing it at 45% for land of average fertility.

(b) The better and more scientifically a field is cultivated and manured, the greater is the increase per cent. On poor soil it is so small as to be scarcely perceptible.

(c) Some vegetables cannot endure the electrical treatment if they are not watered, but then they will give very high percentage increases. Among these are peas, carrots and cabbage.

(d) Electrical treatment, when accompanied by hot sunshine, is damaging to most vegetables, probably to all; wherefore, if favourable results are aimed at, the treatment must be interrupted in the middle of hot and sunny days.

(e) As it is very difficult to determine the effect of electricity on most plants, a special arrangement must be made to get an approximate determination of the increase. In this connection we have already set out (page 60) what must be done in order to avoid the uncertainty which arises from the want of homogeneousness in the soil.

Since the preceding part of this book was written, the following experiments have been carried out during the summers of 1902 and 1903:

In all the experiments mentioned hereafter, the new type of influence machine, described on page 40 (see Fig. 5), and the new insulators were employed. The size of the machine was also the same, and the whole equipment was made at Helsingfors.

ENGLAND

Experiments at Durham College of Science, Newcastle-on-Tyne – Summer, 1902.

After a long correspondence with Mr. Hogarth, of Kirkcaldy, I was induced to erect an installation for experiments to be made at Durham College of Science. After having corresponded with Mr. R. B. Greig, I got the conditions for my experiments settled, and it was determined that they should begin as early as possible. As a matter of fact, they could not begin before 24 May, 1902, after all plants had been sown. This work was completed on 22 May. The machine employed was driven by an electromotor.

Our intention was to put under experiment the following plants: barley, oats, sugar-beets, mangolds, potatoes, beans, carrots, turnips, swedes, clover, rye grass and strawberries.

After the electric current had been applied some days (from 24 May to 2 June), it could be seen on nearly all plants that germination and vegetation were much better on the experimental fields than on the control fields. Especially was this the case with strawberries, and the barley showed a much richer development under current than without it. But after the commencement of June the bad weather—rain and low temperature—began to exercise its strongest influence.

The following meteorological observations, taken at Cockle Park, Morpeth, may be set out to give a true idea of the weather during the summer of 1902:

Period	Hours of sunshine	Days without rain	Average n° of hours machine in motion
24-31 May (7 days)	24.0	4	9.8
1-30 June (30 days)	142.0	14	9.8
1-31 July (31 days)	104.5	10	11.1
1-31 August (31 days)	120.0	12	8.7
1-30 Sept. (30 days)	110.0	14	10.5
1-3 October (3 days)	Rain all the time		

As the rule is certain that the better the vegetation the higher the increase per cent., a good result was not to be expected, in consequence of diminished vegetation generally. But from another cause—viz. that the machine has none or little effect shortly before, during, and shortly after rain—a lowered result must be the consequence. From the majority of plants no harvest could be obtained, because they had not sufficient time to ripen; others were destroyed by insects, and so on. It is therefore to be regarded as somewhat astonishing that the results have been as good as the following table shows:

Plants	Increase		Area of Exp. field = Control field
Strawberries	37.0%		1.49 m
Potatoes (1st)	31.1%	23.3% (average)	8.45 m
Potatoes (2nd)	15.4%		8.45 m
Mangolds	25.0%		8.45 m
Peas	20.0%		8.45 m
Sugar beets	—		4.46 m

A chemical analysis of the sugar-beets made by Mr. S. H. Collins at Durham College of Science gave the following results:

Compounds	Experimental plot	Control plot
Water	74.44%	75.31%
Cane sugar	14.45%	12.24%
Glucose	0.36%	0.38%
Other solids	10.75%	12.07%
Total	100%	100%

Or an increase of 2.21%—or 18.05% of that of the control field—of cane sugar on the experimental field. This analysis agrees fully with an analysis executed in Burgundy in 1888.

One kind of plant, the rye-grass, gave a negative result; but as all the experimental fields were situated lower than the control fields, the former received the water flowing from the latter and, during such a summer as that of 1902, the effect of this was considerable. The control fields had also the advantage of having been more recently manured, and were, in consequence, in a better condition than the experimental fields.

Remarks: The increased percentage of produce on the experimental plots compared with the control plots has here, as always, been determined in the following way:

When the produce on the experimental plots is called *a*, and on the control plots *b*, the formula will be $\dfrac{(a-b)\,100}{b}$ for the increase per cent. When the sign - is put before the figures in the tables, it signifies that the produce on the experimental plot has been less than on the control plot, the formula being, in this case, $\dfrac{(a-b)\,100}{b}$ = decrease per cent.

GERMANY

Experiments at Kryschanowitz, near Breslau, Germany, by Dr. Otto Pringsheim – Summer 1902.

Dr. Pringsheim, who employed a machine of the same size and type as that previously mentioned, obtained the following results from the electrical treatment:

Plants	Increase		Area of Exp. Field = Control field
Strawberries	50.1%		13.4m
Carrots in garden*	13.1%		Number of plants: respectively 775 and 667
Potatoes (1st)	13.8%	20.8% (average)	8.84m
Potatoes (2nd)	17.4%		144.4m
Potatoes (3rd)	30.3%		144.4m
Barley (1st)	6.9%	10.6% (average)	288.0m
Barley (2nd)	14.2%		288.0m
Oats (1st)	40.7%	22.6% (average)	144.0m
Oats (2nd)	4.5%		144.0m

---

* The experimental field was more in shade than the control field. The part of the experimental field which was in shade gave shorter roots.

The machine was driven by an electric motor fed by accumulators, and worked on an area of 1,175 square metres.

The weather was about the same as at Durham College of Science, Newcastle, England, but perhaps a little warmer. The machine was at work about 10 hours, mostly at night. By these experiments it is shown that the abnormal weather has exercised a fatal influence on the vegetation.

SWEDEN
Experiments in Åtvidaberg, by Baron Theodore Adelsvärd – Summer 1902.

– Meslin (barley, oats, peas, &c): an increase of 20.9%.
– Beets (cattle food): an increase of 26.5%.
– Carrots (cattle food): an increase of 2.9%.[19]
– The machine worked on an area of 3.40 ha = 8.4 acres.

As the insulated wire net in these experiments had a great extension, their signification with regard to the use and application of the method is very important; but the rainy and cold weather caused much damage, so that it was almost impossible to obtain sure results. The rain was so abundant that

---

19. Other plants which were under treatment could not be examined, and are, therefore, omitted.

the ground here and there could not absorb all the water, and, in consequence, the vegetation showed an inequality which made the comparison of the two fields very uncertain.

ENGLAND
Experiments at Durham College of Science, Newcastle-on-Tyne – Summer 1903.

The results of these experiments were as under:

Date of harvest 1903	Plants	Crops		Tops or straw	
		Watered	Unwatered	Watered	Unwatered
16 Oct.	Turnips	99.0%	49.5%	99.2%	67.0%
16 Oct.	Sugar beets	40.0%	49.6%	60.3%	66.6%
13 Oct.	Mangels	0.2%	33.2%	20.9%	39.8%
22 Sept.	Rape	27.6%	-20.8%	—	—
8 Sept.	Potatoes	30.8%	65.5%	—	—
8 Sept.	Peas	28.1%	7.9%	85.3%	59.1%
21 Sept.	Beans	-8.3%	-12.0%	16.9%	25.3%
1 Sept.	Rye grass	129.7%	97.4%	—	—
22 Sept.	Clover	13.3%	-16.5%	—	—
4 July to 1 August	Strawberries	-8.3%	38.4%	—	—

Second experimental and control plots gave for potatoes, 26.5%; peas, 9.4%; pea straw, 2.4%; beans, 8.9%; straw, 0.3%.

The machine worked on an area of 100 m^2.

It is shown by this table that turnips, rape, peas, rye grass and clover have been improved by the watering, but not so sugar beets, mangels, potatoes and strawberries, on all of which watering had a marked deleterious influence. Especially was this the case with strawberries, the experimental plot of which had suffered much from the rain in the summer of 1902.

On the unwatered control plot, the harvest of strawberries was, to 20 July, 11.87oz.; on the unwatered electrified plot, on the contrary, 18.25oz., which shows the earlier ripening of the fruits on the electrified plot.

Remarks: In considering the results of these experiments it must be kept in mind that the control field had not been manured within the last four years, while the field in which were the experimental plots had not been manured during the last five years, or a year earlier. The experimental plots also lie about 0.3 m. lower than the control plots, and therefore, may have lost more by washing during the rainy season of 1902 than the higher lying plots. The remainder of the active force of the soil, which evidently existed chiefly at the bottom, was therefore carried away in a higher degree from the experimental than from the control fields. The soil on these fields was mixed in the spring of 1903, but as this could not reach the bottom itself, the experimental squares were in a more unfavourable position than the control squares. The whole field should have been manured in order to produce the best result from the electrical treatment, but the manuring was omitted in order to prevent inequality in the sub-soil.

The soil was too poor to give a high increase per cent., for it is the general rule that the richer the soil and the more luxuriant the vegetation the greater the influence of the electrical current upon the produce.

With the turnips and the sugar beets the surplus of leaves is equal to the surplus of roots, while with the mangold this is not the same, and with the peas and beans the straw and the stalks are far more developed than the fruit itself; this has given a positive result with the peas, but with the beans a negative one. This shows that the soil has become poorer and yet poorer on the experimental fields than on the control fields, and this is confirmed by the circumstance that on the second experimental field potatoes and beans have given negative results and peas a low positive one. This poverty of the soil appears most striking in the reduction of the percentage of sugar, which the chemical analysis shows, concerning sugar beets, to be:

	Control Fields	Experimental fields	Increase	
			Absolute	Relative
Summer 1902	12.24%	14.45%	2.21%	18.1%
Summer 1903	8.71%	9.43%	0.72%	8.3%
Or a reduction of	3.5%	5.0%	—	9.8%

GERMANY

Experiments in Kryschanowitz,
near Breslau – Summer 1903.

Plants		Increase or Decrease	
Strawberries		128.0%	
Sugar Beets	Number 1	119.5%	79.4%
Sugar Beets	Number 2	39.2%	
Barley		32.5%	
Beans		32.1%	
Carrots		-37.6%	
Potatoes		7.6%	

The machine acted upon an area of 517 m^2.

Owing to a misunderstanding, the machine had got a velocity about four times greater than was necessary. This acted badly upon the carrots, which were not watered. This crop has always, under such circumstances, given a negative result. Whether this too great velocity had any noteworthy influence on the other plants cannot be ascertained with certainty. On both the experimental and the control field in the results given under No. 3 (sugar beets), the earth had been mixed, and again laid out in equal quantity on both fields. A chemical analysis of the sugar beets, duly made in a sugar refinery, gave an increase of the percentage of sugar from 14.1 to 15.1, or 7.24% of the whole.

Remarks: This quality of the electric air current (or electrical treatment) of increasing the percentage of sugar in the sugar beets has occurred in the following cases:

Location	Year	Sugar Increase	
		Absolute	Relative
Burgundy	1888	2.69%	18.1%
Durham College	1902	2.21%	18.05%
Durham College	1903	0.72%	8.4%
Åtvidaberg	1903	1.9%	13.4%
Kryschanowitz	1903	1.0%	7.24%

We may regard it, therefore, as sufficiently proved.

SWEDEN

The Experiments in Åtvidaberg – Summer 1903.

We give below a table showing the function of the machine during the different months:

Month	Working		Hours daily	Not Working		Dates
	Number of days	Number of hours		Number of days A	Number of days B	
May (from 23)	9	150	16.7	0	0	
June	26	450	17.3	2	2	1, 2, 3, 18
July	23	369	16.6	1	7	3, 7, 16, 17, 20-23
August	17	311	18.2	8	6	15-17, 23-27
Sept. (until 14)	5	97	19.4	7	2	2, 7-11, 14

Days A: on account of rain.
Days B: from other causes.

From 15 September, no attempt was made to use the machine.

We see by this table that the machine by the month of July begins to show slight irregularities, but that the greatest irregularities occur in the latter part of August and in September. Therefore it seems to serve a practical purpose to divide the plants which were under treatment into two groups. In the first group come corn, rye, barley, oats and meslin, and in the second group the root crops—

potatoes, sugar beets, carrots and beets. The former were reaped in the month of August (oats the 14 September), the latter in the first week of October, except the potatoes, which were reaped in the middle of September. The most important time of development comes, for the former in the months of June, July and August, for the latter in July, August and September. The machine worked very well during the months of May, June and July; but during August and September together for not more than 22 days out of 61, of which 17 days were during August and only five days during September. These circumstances appear clearly from the tabulated results.

For each kind of plants three squares of 16 m^2., situate as shown in Fig. 4 (middle part), page 38, were chosen for the experimental field, and three squares situated in the same way for the control field, within a distance of about 5 m. In each pair of squares (except for the rye, which was autumn corn) the earth from both fields was dug to a depth of 30 cm., mixed, and laid out afresh in equal quantity on each square.

The machine acted upon an area of 4.03 hectares, and the large extension of the wire net in these experiments will give them a particular interest.

The first group—Corn

From Åtvidaberg, I got samples of all species of corn which had been under electrical treatment. There were two samples for each species, one taken from the harvest on the experimental field, the other from the harvest on the control field. The weight of the sample was about 80 g. All the flattened and half grains, and of the oats and barley all those without a shell, having been picked out, these samples had to undergo treatment in a *separator* constructed for the purpose.[20]

The rye was thus divided into three qualities— first, second and third quality—and each had a weight of 1,000 grains. Barley and oats were only in first and second qualities.

The operation was as follows: The corn from the control field was at first separated, the threads of the spiral having been disposed of at a convenient distance from each other. Then the corn from the

---

20. The *separator* consisted of a spiral cut out of a brass tube (12.2 cm. long and 5.7 cm. in diameter), which could be shortened or lengthened. Through this, the distance between the threads of the spiral could be enlarged or diminished as required. The threads of the spiral had a breadth of 0.36 cm. and a thickness of 0.18 cm., and their edges were kept sharp. The spiral was furnished with bottom ends with circular holes in the middle, and was, when in a horizontal position, pushed on to the axis of a centrifugal machine, which was kept by an electric motor in slow rotation (1 revolution in 2 seconds) to avoid the effect of centrifugal force.

experimental field was separated, the distance between the threads of the spiral being the same.

Rye

From the field sown in the autumn of 1902, squares were taken for the experimental field as well as for the control field. The harvest, calculated by the hectare (2.47 acres) was:

	Crop kg.	Straw kg.	1st Quality.		2nd Quality.		3rd Quality.	
			Per cent.	Weight of 1,000 grains in g.	Per cent.	Weight of 1,000 grains in g.	Per cent.	Weight of 1,000 grains in g.
Exp. field...	1,917	5,583	47·6	28·85	27·8	22·03	24·5	16·38
Control field	1,604	4,646	38·3	27·04	32·5	21·72	29·2	16·04
Increase ...	19.5%	20·2%	9·3	1·81	—	—	—	—

The electric air currents had thus augmented the rye harvest by 19.5%, and improved the quality of the corn so that we obtained of first quality 9.3% more on the experimental field, or under the current, than on the control field. But, besides that, the weight of 1,000 grains had augmented by 1.81 g. on 27.04, or by 6.7%.

According to the statement of the Manager, Mr. Tillberg, the yearly average of produce per hectare for the three last years was in Åtvidaberg 2,347 kg., and was this year 31.7%, lower than this average. This circumstance also caused a diminishing of the increase per cent., according to the rule already mentioned that "the better the vegetation the higher the increase per cent."

Barley

This crop was sown on the experimental and control fields on which the soil had been, as above mentioned, mixed. Calculated per hectare, there was obtained:

	Crop kg.	Straw kg.	1st Quality.		2nd Quality.	
			Per cent.	Weight of 1,000 grains in g.	Per cent.	Weight of 1,000 grains in g.
Exp. Field ...	3,708	7,229	47·9	49·64	52·1	38·99
Control Field	2,646	4,021	42·6	49·59	57·4	39·33
Increase	40·1%	79·8%	5·3	0·05	—	—

The barley had thus given a surplus crop of 40.1% and a 5.3% increase of first quality on the experimental as against the control field. The surplus weight of 1,000 grains is here only 0.05 g.

Oats

The experimental and control fields were treated in the same way as for barley. The harvest, calculated per hectare, was:

	Crop kg.	Straw kg.	1st Quality.		2nd Quality.	
			Per cent.	Weight of 1,000 grains in g.	Per cent.	Weight of 1,000 grains in g.
Exp. Field ...	2,271	2,521	76·5	50·20	23·5	27·16
Control Field	1,258	2,521	67·9	49·45	32·1	29·44
Increase	16·0%	0·0%	8·6	0·75	—	—

The harvest of oats shows the singularity that the harvest of straw is the same on both fields, although there is an increase of crop of 16.0%. For the other species of corn, the increase of crop is accompanied more or less by an increase in straw. The explanation partly lies in the improvement of the crop which the electric air current had caused (the percentage of first quality on the experimental field exceeding that on the control field by 8.6%, and also the weight of 1,000 grains on the former field over that of 1,000 grains on the latter by 0.75 g.); but this does not seem sufficient to explain the above-mentioned singularity. We must therefore suppose that the number of grains of oats on each straw has been greater on the experimental field. A fault in the drainage of this field was observed during the summer, which had reduced the fertility of the field, and thus a falling off of the increase per cent.

Meslin

This consisted of barley and oats, and occupied the greatest part of the field, or 2.52 hectares

(= 5.06 acres). It was observed that the net over this field became, from time to time, in contact with the plants, and had to be carefully looked after in this respect. During these interruptions the current was stopped; thus the current acted for a shorter time on this field. Another cause of the shortening of time was that the net was twice destroyed by straying horses. On the first occasion, the damage was repaired in nine days, so that the current was not operating upon this field from 31 May to 11 June (in all 12 days, of which 3 days was from another cause). The second time there was no current from 29 June to 1 July.

For the meslin, there were two experimental fields and two control fields, of the same size and quality as for barley and oats; and again, on these fields the soil was mixed. The samples were taken from both fields, which we will call A and B. For the purpose of examining the corn, the oats were separated from the barley, and showed the following proportions:

	A		B	
	Exp. Field. Per cent.	Cont. Field Per cent.	Exp. Field. Per cent.	Cont. Field. Per cent.
Barley	64·5	57·8	34·1	33·0
Oats	35·5	42·2	65·9	67·0

From this, we see that the barley on A represented about two thirds and on B one third, and the oats one third and two thirds respectively. After deciding the proportion between barley and oats, the corn was cleaned in the way previously mentioned. The harvest from the first fields was:

	Crop. kg.	Straw. kg.	1st Quality.		2nd Quality.		
—			Per cent.	Weight of 1,000 grains in g.	Per cent.	Weight of 1,000 grains in g.	
Exp. Field A	2,583	3,453	53·8	57·16	46·2	46·00	} Barley.
Cont. ,,	1,812	2,042	48·8	56·92	51·2	43·02	
Exp. ,,	—	—	88·2	45·39	11·8	17·88	} Oats.
Cont. ,,	—	—	87·8	46·85	12·2	17·84	
Increase	42·6%	69·4%	5·0	1·24	—	—	Barley.
,, ...	42·6%	69·4%	0·4	−1·4	—	—	Oats.

The meslin has thus given an increase of 42.6% for barley and oats together; but whilst the barley shows a surplus of 5% on first quality, the surplus for oats is only 0.4%, which we may regard as equal zero; and while barley, for a weight of 1,000 grains, gives a surplus of 1.24 g., this is reduced for oats to a negative of -1.4 g.

On the experimental field B, the harvest was as follows:

	Crop. kg.	Straw. kg.	1st Quality Corn.		2nd Quality Corn.		
			Per cent.	Weight of 1,000 grains in g.	Per cent.	Weight of 1,000 grains in g.	
Exp. Field B	2,313	3,937	51·7	56·47	48·3	44·70	Barley.
Cont. ,,	2,500	3,542	43·4	54·92	56·6	43·77	
Exp. ,,	—	—	80·4	46·46	19·6	21·87	Oats.
Cont. ,,	—	—	80·6	46·38	19·4	25·12	
Increase	−7·5%	11·2%	8·3	1·55	—	—	Barley.
,, ...	—	—	−0·2	0·08	—	—	Oats.

The barley on the experimental field B thus gave a surplus of 8.3% first quality and of 1.55 g. in the weight of 1,000 grains, whereas the corresponding quantities for oats were -0.2 and 0.08, or almost equal to zero. Oats, which, when cultivated alone, show so important an improvement as 8.6% increase in first quality and 0.75 g. in the weight of 1,000 grains, show, when cultivated along with barley under electrical treatment, no such improvement, which is a most remarkable result. Probably the cause is that the barley has grown taller than the oats, and thus received more of the electric current. The whole quantity of oats is, nevertheless, increased in the same proportion as the barley.

These two second experimental and control fields give a negative result of -7.5% in crop, but 11.2% surplus in straw. The cause of this lies in the exceptionally large harvest which the second control field has given, this surpassing the harvest from the first control field by 688 kg., or about 38%, and the yearly average harvest of meslin, which during the last three years had been 2,047 kg. per hectare, by 453 kg., or 22.1%. This unusual fertility seems to depend on chance, and therefore the results of the second experimental and control fields are wholly excluded from the account, especially as the barley on this experimental field shows the same improved qualities as that on the first experimental field.

We must now call attention to the difficulties which always arise in the choice of experimental and control fields. Already at these experiments, which embrace only four different cornfields, accidental circumstances have, in two cases, diminished the increase per cent., or have transformed it into a minus quantity. In the first case, it was a fault in the drainage, in the second case, an unusually great fertility of a control field; and this, notwithstanding the previous careful mixing of the soil. Unfortunately, the quality of the subsoil seems to be of such great consequence to the success of the experiments that this method has not been a sufficient safeguard. It will therefore be necessary to improve the method to ensure a regular increase per cent. (See pages 60 and 66).

If all circumstances are taken into consideration, we must conclude that the increase per cent. during 1903 has been between 35 and 40%, for the corn crop, added to the improved quality of the corn, shown not only by the Tables above, but especially by the chemical analysis of the crops (page 103).

The second group—Roots

Crop.	Increase per cent.	On the Exp. Field.		On the Control Field.	
		No. of Roots.	Mean Weight of One Root in g.	No. of Roots.	Mean Weight of One Root in g.
Potatoes	4·8	...	...	...	...
Sugar beets ...	6·22*	398	357	406	337
Fodder beets	−2·4	305	803	316	823
Carrots	−8·9	578	358	549	393

A chemical analysis of the sugar beets gave as the percentage of sugar of the beets from the experimental field an increase of 1.9%, or from 14.2%, to 16.1%, which makes 13.4% increase on the percentage of sugar in the sugar beets from the control field.

The increase percentages are small for potatoes as well as for sugar beets, and it comes near the fact to suspect the cause to be a greater fertility of the control fields, in analogy with the second experimental and control fields in the case of the meslin. It is known with certainty that, on the mixing of the soil, the subsoil of the fields showed very

---

* The increase per cent. is determined by the mean weight of one root.

different constituents. In favour of this view also, there is the negative result with the fodder beets, which always previously have given a fairly high increase per cent. If the fertility of the two fields had been the same, and if the electric air-current had had no effect, the harvest ought to have been the same from both fields, but the negative result shows a smaller fertility of the soil of the experimental field. Carrots, which have always been a delicate crop, cannot here serve as a proof.

At all events, the summer was so wet in the latter half of August and the earlier part of September that the condition of the subsoil had a marked influence on the fertility of the soil. That the electric current had exercised an influence on the roots during a long period, and ought thus to have produced a higher increase per cent., if both fields had been of an equal fertility, is shown by the great increase of the percentage of sugar in the sugar beets, for such a condition is not produced suddenly.

The chief cause, however, of the low increase per cent. lies in the circumstance that the influence machine from 1 August to 15 September had not been at work for more than 22 days; hence there were 23 days of inactivity. From 15 September to 8 October, or 23 days further, the machine was out of use owing to a misunderstanding of the skilled manager in Åtvidaberg, Mr. Tillberg, who, with never-failing interest, exercised the supervision and direction of these experiments. He reported on the 15 September that, because only sugar beets, fodder beets and carrots were unripened, he did not think it worthwhile to let the machine work. There were in all 46 days of inactivity, among them 15 rainy days, during which it was indifferent whether the machine worked or not. Altogether, therefore, we have 38 days when the machine was out of use. There was a further eight days of inactivity, three being in consequence of necessary repairs and three days because of great conducting power of the air due to its humidity.

During two of the eight days, this humidity of the atmosphere caused another difficulty, due to *radio-activity*. This quality of the air was discovered by Elster and Geitel in 1902-1903, and acts chiefly in such a manner that the air has a great conducting power for electricity. When this condition of the air arrives, the only thing to do is to wait for its cessation, which happens generally after some few hours, or at most one day. During June and July, it occurred but three times. Mr. Tillberg tried to neutralise the effect of this radio-activity by putting new cylinders on the machine and by varnishing the old cylinders, &c. This work might have been avoided, however, for, after having ascertained conclusively that the air in the room is quite dry, and after having given the cylinders a careful drying by warming them, we can but wait for the cessation of the conducting power in the air before resuming operations.

According to information from Åtvidaberg, the weather during the latter part of September was not actually wet, but humid. It is, therefore, probable that the machine could have been kept in activity during 14 days more. Remembering that the effect of the electric air current is nearly the same as that of sunlight, it is very likely that a continuance of the electrical air current treatment during these two weeks would have augmented to a perceptible degree the increase per cent. of the roots. One consequence of the experiments this year is evident, namely, the necessity of improving the method of control. (See pages 60-61.)

We will now make a comparison between some of the plants which on the different places have been under treatment during 1903.

Place.	Increase per cent.			Increase per cent. of sugar in sugar beets.	
	Barley.	Sugarbeets.	Potatoes.	Absolute.	Relative.
	Per cent.	Per cent.	Per cent.	Per cent.	Per cent.
Kryschanowitz	32·5	79·7	7·6	1·0	7·2
Durham College, England	...	49·6	65·5	0·7	8·4
Åtvidaberg ...	40·1	6·2	4·8	1·9	13·4

As the machine has worked in these places on very different areas, we could understand if the increase per cent. had a relation to the area; but this is not the case, for in Kryschanowitz, near Breslau, where the machine worked on an area of 517 m^2., there was an increase per cent. of barley of 32.5%, whilst in Åtvidaberg, where the machine worked on an area of 40,300 m^2. (4.03 hectares), the same crop gave 40.1%, or 76% more. We must therefore conclude that the wet weather which prevailed in Åtvidaberg during August and September caused this small increase of the roots, and also, that a greater fertility prevailed there on the control fields for the same plants. This remarkable fact is also observable in the harvest of potatoes. In Kryschanowitz there was 7.6%, and in Åtvidaberg 4.8%; but at Durham College, where the machine worked on an area of only 100 m^2., the increase was 65.5%. The machine at Kryschanowitz had a speed of 3.5 turns per second, at Åtvidaberg 2.5 turns, and at Durham College about 1.0 turns; the quantity of electricity generated, and therefore distributed, is almost proportional to the number of turns per second.

The main cause contributing to this unequal result is, without doubt, to be found in the inequality of the soil of the experimental and control fields, and therefore it is necessary to improve the method of controlling, as previously stated.

If all the above enumerated conditions had been duly attended to during the experiments, it is very probable that the increase per cent. for the roots would have been about the same as for the crops, in addition to the improvement in the quality of the roots.

It is a great pleasure to me to express to the professors and teachers at Durham College of Science, England, my best thanks for all their kindness in assisting me, but especially to Mr. R. B. Greig, who has given me so much of his valuable time; and to Dr. Thornton, who helped me with the arrangement of the electric motor and its adaptation for my purposes. Mr. Greig having been transferred to the University of Aberdeen, Prof. Gilchrist had the kindness to allow the experiments to be continued, and Mr. Bryner Jones undertook the onerous task of supervising the experiments and sending me the results. For all this, I beg to express to them my best thanks.

To Dr. Pringsheim, to whom I stand in such great obligation, not only for the indefatigable care he devoted to the experiments and to their carrying out at Breslau, but also for the great kindness and hospitality shown to me by Mr. and Mrs. Pringsheim during my stay in Breslau, and particularly at Kryschanowitz, I am very happy to express my sincerest thanks.

It is also a pleasant duty to me to express my deep thankfulness to Baron Adelsvärd for his goodwill and assistance in placing his land at my disposal for the experiments. I am all the more grateful to Baron Adelsvärd, as he gave me a conscientious and active assistant in the person of his skilled and interested manager, Mr. Knut Tillberg, who carried out and supervised in an able manner the experiments, and also provided me with all necessary workmen for my purpose.

CHEMICAL ANALYSIS OF CROPS AT ÅTVIDABERG WITH AND WITHOUT ELECTRICAL TREATMENT.

[NOTE.—The increase per cent. of a substance is calculated after the formula $\frac{(a-b)100}{b}$, where a represents the quantity which was received under electric current, and b the quantity received outside the electrical current.

SUBSTANCE.	RYE.			BARLEY.			OATS.			OATS separated fm. meslin.		
	Under electric current.	Outside electric current.	Incr'se per cent.	Under electric current.	Outside electric current.	Incr'se per cent.	Under electric current.	Outside electric current.	Incr'se per cent.	Under electric current.	Outside electric current.	Inc. or dec. percent
Mineral substance (ash)	2·128	2·440	—	2·979	2·950	—	3·712	3·739	—	3·899	4·107	—
Ether-extract (raw fat)	2·516	1·799	—	2·018	2·443	—	4·831	5·358	—	5·255	5·339	—
Cellulose	2·296	2·379	—	5·605	5·015	—	11·136	9·434	—	9·944	11·655	—
Nitrogen free extract	81·730	83·922	—	77·158	77·502	—	67·781	69·429	—	69·002	66·853	—
Raw proteid matter	11·270	9·460	19·13	12·240	12·090	12·41	12·540	12·040	4·15	11·900	12·040	−1·16
TOTAL	100	100		100	100		100	100		100	100	
Per cent. of Nitrogen:												
Total nitrogen	1·803	1·513	19·17	1·958	1·935	1·19	2·007	1·927	4·15	1·904	1·927	−1·19
Amide nitrogen	0·218	0·110	—	0·094	0·061	—	0·093	0·112	—	0·086	0·131	—
Albuminoid nitrogen	1·427	1·249	14·25	1·507	1·599	2·00	1·795	1·660	8·13	1·724	1·661	3·8
Total digestible nitrogen	1·645	1·359	21·05	1·661	1·660	0·06	1·888	1·772	6·55	1·792	1·792	0·0
Non-digestible nuclein	0·148	0·154	—	0·297	0·285	—	0·119	0·155	—	0·112	0·135	—
Digestibility of nitrogen	91·8	89·8	—	84·8	85·8	—	94·0	91·1	—	94·1	93·0	—
In 100 parts of Nitrogenous Matter:												
Amides	12·7	7·2	—	4·8	3·2	—	4·6	5·8	—	3·6	6·8	—
Digestible protein	79·1	82·6	—	80·0	82·6	—	89·4	86·1	—	90·5	86·2	—
Non-digestible protein	8·2	10·2	—	15·2	14·2	—	6·0	8·1	—	5·9	7·0	—

From this analysis it follows that the electrical air current has produced an increase of the proteid matter in the rye of 19· per cent., in the total quantity of nitrogen 19·2 per cent., in the albumen of 14·3 per cent., and in the digestible nitrogen of 21·1 per cent. In the barley the electric air current only produced in the raw proteid matter an increase of 12·4 per cent., in the total quantity of nitrogen 1·2 per cent. In the oats, on the contrary, it has produced an increase of 4·2 per cent. of proteid matter, of 8·1 per cent. in the albumen, and of 6·6 per cent. in the digestible nitrogen; whereas in the oats separated from the meslin an increase of albumen of only 3·8 per cent. was produced. All crops under the electrical treatment had been improved in quality to a considerable extent. For rye this improvement can be taken at 20 per cent., for barley at 12 per cent., and for oats at 10 to 12 per cent.

The above analysis was made in the Agriculture Economic Laboratory at Helsingfors (under the direction of Prof. Arthur Rindell) by candidate of philosophy Mrs. Lilly Wendt.

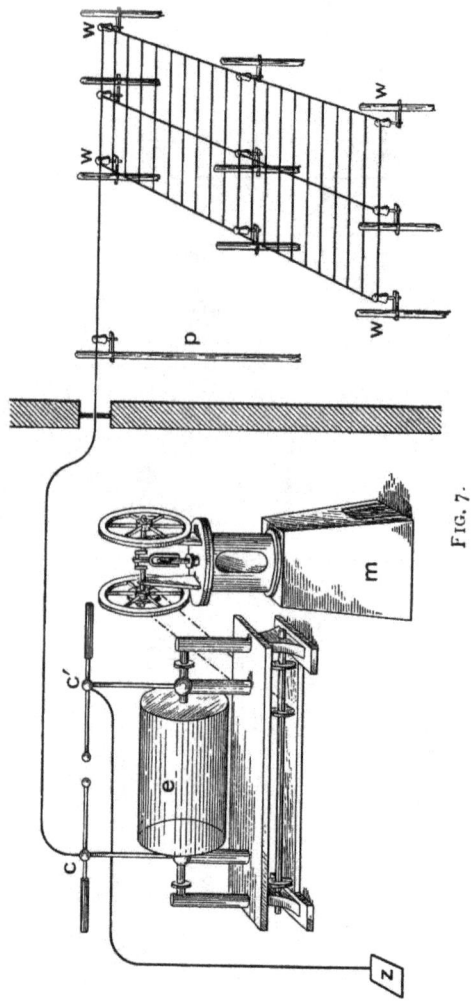

Fig. 7.

RULES FOR APPLYING THE ELECTRICAL AIR-CURRENT ON GROWING PLANTS

Introduction: The principal manner in which electricity, or an electrical air current, exercises its influence on growing vegetables has been discovered to consist (1) partly in producing ozone and nitric compounds and introducing them into the capillary tubes of the plants when the current is going from the points of the wire net to the earth (positive); and (2) partly in drawing up the sap from the roots upwards when the current is going from the earth to the points (negative),[21] the former effect being about 30% greater than the latter. If we suppose that the former or positive current is applied, it will be easy to apply it for the purpose, and to give clear rules for this application. The lengthy experience I have had in these experiments enables me to give these rules, which, if they are intelligently applied, will lead to favourable results.

I.

The whole scheme of the experiments will be easily understood from the diagram (Fig. 7) on the opposite page.

---

21. See *On the State of Liquids in Capillary Tubes under the Influence of an Electrical Air-Current*, by the Author. (See p. 62).

From the positive pole c' of the influence machine is conducted an insulated copper wire through a hole in an ebonite disc fixed in the wall, over the posts p to the insulated wire net $w\ w\ w\ w$ over the growing plants. The negative pole c is conducted to a zinc plate, z, in humid soil. Following are details of:

1. The electric influence machine, its installation, and the motor.
2. The experimental and control fields.
3. Putting the wire net on the insulators and their fixing.
4. Remarks.
5. Cost of the experiments.

II.

The Electric Influence Machine and its Installation.— As the old machines of Holtz and Wimshurst have shown, especially on prolonged use, many inconveniences, I have constructed the type represented in Fig. 5 (page 40), in which these inconveniences are, for the most part, removed. This machine consists of two cylinders of glass or ebonite, E, rotating in opposite directions, one within the other, and provided with metallic (tin) boats fixed on strips of tinfoil. Over the machine is placed a cover fitted with a drying apparatus not shown in the figure. The glass cylinders are covered with a special varnish to protect them from humidity, and the brushes are made of silver-thread.

The machine is best charged if first put in rotation by some turns in the wrong direction, the outer cylinder going from the brushes of the transverse conductor against the points of the collector, and afterwards in the right direction—that is to say, the outer cylinder rotating from the collector against the brushes, causing sparks to go between the separate spheres of the discharger. The spheres must be separated of at least 2 cm., and some sparks taken before the machine comes to full charge. In experiments on the land the machine must be installed in a dry room, preferably on an upper floor, but always some height over the soil. In this room, the temperature must be kept always some degrees higher than the outside. In such places, the drying apparatus referred to above is not necessary, but the cover must always be kept over the machine to protect it from dust.

The machine charging the positive pole must be connected to the insulated wire net by means of insulated copper wire, and the negative pole with a plate of zinc (0.5 m^2.) in humid soil.

The method of distinguishing the negative from the positive pole is simple. Take away the connection of the Leyden jars with the poles on the one side and also the conducting wire (to earth or the insulated wire net), and put the spheres of the discharger at a distance apart of about 1.5 cm. If the machine is moved, a stream of light is seen between the spheres, violet on the side of the negative pole, with a small bright spot on the sphere, white and very bright, on the positive side. The machine is easily made to perform this experiment. As the machine, when stopped, very frequently changes its poles, this experiment must always be repeated and the machine continue in motion before fixing the conducting wires.

The new machine does not give so long a spark as the older one, but the quantity of electricity generated and given out is much greater. If a machine of the new type, of medium size, with glass cylinders of 40 cm. length and about 30 cm. diameter, gives 100, the Wimshurst, with plates of 45 cm. diameter, gives only 27, and in some cases only 22; while the old Holtz machine, with four plates of 46 cm., gives 30. But the greatest advantage of this new type of machine is that it can be kept in motion for three months if necessary, cleaning the glass cylinders between the tin strips, with a linen cloth, only being necessary from time to time. After long use the glass, cylinders must be thoroughly cleaned up and fitted afresh with new tin strips and boats, if necessary, and be also covered with new varnish.

The Motor.— Any small motor.—electric, hot-air or otherwise—is suitable to do the little work required to move the machine. If an electric motor is used it can be battery-driven. Small hot-air motors are suitable for this work. The machine, b, shown in the illustration (Fig. 7) has an output of one tenth of a horse-power and costs about £8 sterling, or, say, $41. It is easily managed.

The Field.— If the purpose is to apply the electricity in such a way that the results can be brought clearly forward for comparing the yield on two fields, of which the one is under electrical treatment (experimental field), the other serving as the control field, we have to observe the following points:

(a) CHOICE OF FIELD

The qualities of the soil must, as far as possible, be the same throughout the whole field—of the same composition, the same humidity, the same fertility, and of about the same height over a common level. The field must, further, be in the same way exposed to the sun, so that no shadow at all (*eg.* from growing trees) falls upon it. We will suppose that the experiments are to be made on six kinds of plants, and we then divide the field into squares, as shown in Fig. 8:

E_1	E_2		C_1	C_2
E_2	E_1		C_2	C_1
E_1	E_2		C_1	C_2
E_3	E_4		C_3	C_4
E_4	E_3		C_4	C_3
E_3	E_4		C_3	C_4
E_5	E_6		C_5	C_6
E_6	E_5		C_6	C_5
E_5	E_6		C_5	C_6

FIG. 8.

The side of every square is 5 m^2., making the area 25 sq. m. When the experiments are made in a garden, the squares can be arranged in two beds or layers, with a pathway between them. Such paths must be made between the ranges of squares. In Fig. 8 the squares marked E1, E2, &c, represent experimental fields, and the squares marked C1, C2, &c, the corresponding control fields. Between the two ranges of fields is left a space of 5 m., which can be sown with any sort of plant, but must be out of consideration so far as the experiment is concerned.

If we number the kinds of plants with which experiments are to be made,

        1   2   3   4   5   6,

it is safest to sow them as indicated in the figure.

The plants No. 1 are sown where we have E1 and C1, &c. In such an arrangement we shall get three experimental fields and three control fields for every sort. If there is any inequality in the soil it will, for the most part, be eliminated by the results from the three fields.[22]

The seed must, of course, be chosen with much care, and must be as equal as possible for both kinds of fields.

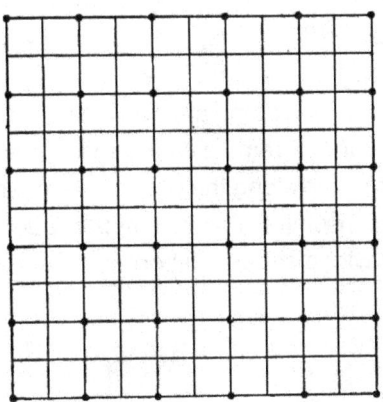

Fig. 9.

If a general application of electricity is to be given, the posts are to be fixed as shown in Fig. 9. In this figure the round dots signify posts provided with insulators, as described below. Along these insulators the wire is carried.

---

22. *See* the remarks made on page 62.

III.

Fixing the Wire Net and Insulators: Along each range of the experimental squares posts, 2.5 m. in length and about 6 cm. in diameter, are rammed in the ground very firmly to a depth of 1/2 m. and at a distance apart of 14.3 m. Every corner-post must have on the inner side (the side of the experimental field) a short stout support to resist the strain of the iron wire. The insulators (Fig. l0) have now to be firmly fixed to the posts, putting the ring *b* around the post and drawing the screw.

When the insulators are well fixed, galvanised iron wire (of 1.5 mm. diameter) is carried with a medium tension along the posts, over the porcelain bell of the insulator. The wire is put round the bell and fixed with a separate short wire.

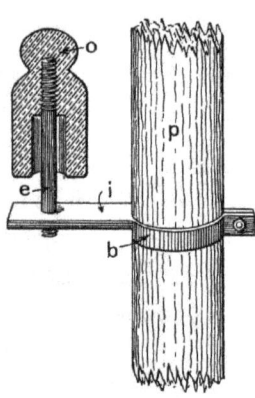

FIG. 10.—*b*, ring; *e*, ebonite tube; *i*, iron bar; *o*, porcelain bell; *p*, post.

When these conducting wires have been fixed all round the experimental field, cross-wires of about 0.6mm diameter are laid on them from one side to the other at a distance of about 1.25 m., and in this way, eight cross-wires cover 10 m. This finer wire must be furnished, at 1 m. apart, with small points of 2 cm. in length, nearly similar in appearance to the well-known barbed wire.

The wire net, being now fully insulated from the posts, is, as stated above, connected by means of a well insulated conducting wire to the positive pole of the electric machine, the negative pole being conducted to earth by means of a wire and a zinc plate. When the wire between the net and the machine is led through a wall, it must be insulated with a round plate of ebonite (4 cm. diameter), as shown in Fig. 7.

After some sparks have been taken between the spheres of the discharger, the spheres are to be put in contact and the wires connected with c and c'. Here, after the spheres are separated, the wire net is charged. To prove this, the spheres of the discharger are brought to within a short distance of one another (1 mm. or less), and a small spark will be seen between them, notifying that the net is charged, well insulated, and fitted for its purpose.

The height of the wire net above the plants must be about 0.4 m. at its lowest point, and this height must be increased as the plants are growing. This raising needs to be done only once or twice, by loosening the rings of the insulators, lifting them up and again fixing them.

The wire net must be carefully examined to see that it does not touch the plants, as all the electric current will be conducted away by such contact.

As the electricity, if given in too large quantity, may damage the plants, especially if administered at periods of burning sunshine, it is best to apply the current on sunny days for only four hours in the morning (from 6 am or 7 am to 10 am or 11 am), and for four hours in the afternoon (from 4 pm or 5 pm to 8 pm or 9 pm).

On cloudy days, however, it can be given the whole day, and even during the night when the weather is moist. *During rainy days, it is useless to keep the machine in motion, because the wire net loses its charge instantly.*

In very dry periods, with burning sun, it is best to interrupt the applications of the electricity, or to administer for only one or two hours a day.

When the air is very moist, the machine will speedily lose its charge, because the electricity is going from the wire net almost instantly. On such occasions, it will be advisable to put the spheres of the discharger so near that sparks are intermittently passing between them. These sparks are very small, but they will keep the machine charged; a

Leyden (Lane)[23] jar in the conductor to earth will have the same effect.

Remarks: During the experiments with this installation attention must always be directed to the insulation of the wire net and the charging of the machine.

The insulators must be carefully observed every day, because spiders make their webs over from the wire net to the posts. This web must be cleared away.

When the electric machine is kept in a dry room, it will be charged at the first turns. In the summer, it sometimes happens that the air in the room is colder than that outside. In this case, a condensation of watery vapour begins within the machine, and its insulation is destroyed. (This has not hitherto been observed, because the older machines have never been used for continuous work, and have always been kept in an ordinary room with dry air.) The best course is, therefore, to put the machine, together with the motor, in a small room on an upper floor (about 2 square metres in area), provided with a small fireplace with the fire laid ready to be lighted, or a small heating apparatus with a source of heat outside, care being taken that the temperature inside the room is always two or three degrees higher than that outside.

---

23. A Leyden jar transformed into a Lane jar is a jar with which the length and the number of sparks can be measured.

The best method to observe, with regard to the supply of the electric current, is to construct, as it were, a central supply station, to branch out conductors 5km on all sides around it, and to sell the electricity as an ordinary article of commerce. In this way, the cost of current will be ascertainable, and this is most desirable from all points of view.

COST OF THE INSTALLATION

The cost of an installation for the application of electricity to growing plants on a field of an area of 10 hectares (24.7 acres), or on 1 hectare (2.47 acres), would be as follows:

	For 10 Hectares.	For 1 Hectare.
Electrical Machine and Motor—		
One influence-machine of medium size, with reverse rotating glass cylinders of about 30 cm. (11¾in.) in diameter and 40 cm. (15¾in.) in length	£32 0 0	£32 0 0
One hot-air motor, one-tenth horse power	10 0 0	10 0 0
Insulators, Wires and Posts—		
640 insulators for the field and 10 for the conductor to the field (at 1s. 1d each)	35 4 2	3 13 9
13,800 metres galvanised iron wire of 1·5 mm. in diam. (weighing 13·4 kg. per 1,000 m., equal to 18·5 kg.), at 7·7d.	5 18 8	0 11 11
81,000 m. galvanised iron wire (1,000 m. weighing 2·48 kg., equal to 201 kg.) at 11½d.	9 12 8	0 19 3
Labour for placing points on this wire (150 m. in one hour equals 540 hours at 4d.)	9 0 0	0 18 0
640 posts 2·5 m. in length and 7 cm. in diam.	2 11 3	0 5 2
10 posts 5 m. in length and 8 cm. to 10 cm. in diam.	0 10 0	0 5 0
Installation of machinery, motor, &c.	4 0 0	4 0 0
Say,	£108 16 9	£52 13 1

	For 10 Hectares.	For 1 Hectare.
Annual Charges —		
Setting up and taking down wire net each season	£3 6 8	£1 0 0
Inspection and supervision (2 hours daily)	5 0 0	1 0 0
Contingencies, renewals, &c.	4 0 0	1 0 0
Interest, 10% on capital, £108 16 9	10 16 10	5 5 3
	£23 3 6	£8 5 3

REVENUE

We will take the wheat as a crop at medium price to estimate the revenue from the application of the electrical air-current. A field sown with wheat gives, speaking generally, 34 bushels per acre as an average crop on ordinary good wheat land; then an increase of 45% on 25 acres, or say 10 hectares, is equal to 383 bushels.

383 bushels at 3 s. 6 d. per bushel equals:

	£67	0	6
Less the yearly cost:	£23	3	6
The net profit on 24 acres equals:	£43	17	0

Or more than four-tenths of the whole cost of the complete installation will be realised in the first year.

As the cost of the installation does not increase in direct proportion to the area of land treated, it follows that on larger areas the profits would be greater, whereas on a very small area (1 or 2 acres) the increased crop would not pay for the application of the electric current.

In the case of horticulture, however, it is an entirely different matter. The produce has a much higher value, and the electrical treatment will always repay the cost when the area is not too small and is very intensively cultivated. In this department of agricultural work, therefore, the future for the electrical treatment of growing crops has a most promising future.

Remarks: In other countries, where the wheat prices are much higher, this process of calculation will give, generally speaking, 30 hectolitres[24] per hectare, or 300 hectolitres for 10 hectares. In these countries also the capital cost and maintenance charges are, speaking generally, lower than in the United Kingdom. The electrical treatment gives an increase of 45%. Therefore:

135 hectolitres at 15.3 s. per hectolitre, equals:	£103 5 6
Less the yearly cost:	-23 3 6
The net profit on 10 ha will be:	£80 2 0

or nearly three-quarters of the whole cost of the complete installation. This great difference comes from the very low price of wheat in Great Britain. The price per bushel varies very much:

—	New York	Berlin	Stockholm	Helsingfors
	Price	Price	Price	Price
Hectolitre	12s 0d	13s 4d	12s 7d to 15s 8d	17s 1d* to 18s 11d**
Bushels	7s 5d	7s 11d	5s 8d to 7s 6d	6s 3d* to 6s 10d**

* American wheat.   ** Russian wheat.

---

24. One hectolitre equals 2.751 bushels.

Though the above calculations on the financial side of the question are merely approximate, there will undoubtedly be great pecuniary advantage in applying the electrical treatment, when we take into consideration not only the increase in the harvest but also the improvement of the crops, as shown by the chemical analyses. That the electrical air-current will prove of great service in orchards and on large market growing lands, by inducing earliness and rapid maturity as well as increased crops, seems to be clearly demonstrated.

The cost of the installation includes the electrical machine, which is put down for a sum of £32, or $164, and this sum may be regarded as very high in comparison with the cost of the older machines. But my own experience will, here, be useful. He who would avoid the troubles I experienced in working with the older machine will not hesitate to use the new type, with its opposite-direction rotating cylinders. This new type of influence machine has the following advantages in comparison with the older ones:

The New Machine.	*The Older Machine.*
(Glass cylinders 40 cm. in length and 30 cm. in diam.)	(45 cm. plates.)
1. It never fails to be charged.*	1. The older machine fails very often to be charged, especially in the summer, when the air is sultry and humid.
2. It can be kept at work during several months continuously.	2. When at work it must be cleaned every third day.
3. It can be used on a greater area (say up to 10 hectares).	3. It can only be used upon an area not exceeding 2.5 hectares.

These advantages of the new machine are so great, especially No. 1, that the superiority of this new type of apparatus is fully proved, and its use justified.

To prevent mistakes and subsequent trouble, it will always be best to order the machines and the necessary tools from an actual manufacturer, and to have the installation made by skilled workmen. Everything naturally depends, in so new an application, upon the skilful first installation and the subsequent careful use of the mechanism and the materials employed.

---

* Excepting when the air is *radio-active*, which will occur very seldom, say four or five times in the summer, and then only for a few hours.

LINEAL MEASURE WITH VALUES IN INCHES.

10 millimetres	= 1 centimetre	= 0·3937079 in.
10 centimetres	= 1 decimetre	= 3·937079 in.
10 decimetres	= 1 metre	= 39·37079 in.
10 metres	= 1 decametre	= 393·7079 in.
10 decametres	= 1 hectometer	= 3,937·079 in.
10 hectometres	= 1 kilometre	= 3,9370·79 in.
10 kilometres	= 1 myriametre	= 393,707·9 in.

A metre is equal to ..
- 39·37079 inches
- 3·2809 feet
- 1·0936 yards
- 0·1988 pole
- 0·04971 chain
- 0·004971 furlong
- 0·0006213 mile

Proportion to Metre.

				M.	Fur.	Yd.	Ft.	In.
Millimetre	1/1000		0·03937079 inch					about 1/25
Centimetre	1/100		0·3937079 inch					1/25 nearly
Decimetre	1/10		3·937079 inches					
Metre	1						3	3·37079
Decametre	10		32·80899 feet				10	2	9·7079
Hectometre	100		109·3633 yards				109	1	1·079
Kilometre	1,000		1,093·633 yards			4	213	1	10·79
Myriametre	10,000		6·21382 miles	6	1	156	0	11·9

The land chain is 1 decametre in length, and is divided into 50 links, each 2 decimetres in length. Sometimes the double decametre is used, which nearly equals the English land chain. Long distances are expressed in kilometres.

Metres into Yards.		Kilometres to Miles and Yards.			Metres into Yards.		Kilometers to Miles and Yards.		
Metres.	Yards.	Kilo-metres.	Miles.	Yards.	Metres.	Yards.	Kilo-metres.	Miles.	Yards.
1	1·094	1	0	1,094	20	21·873	20	12	753
2	2·187	2	1	427	30	32·809	30	18	1,129
3	3·281	3	1	1,521	40	43·745	40	24	1,505
4	4·374	4	2	855	50	54·682	50	31	122
5	5·468	5	3	188	60	65·618	60	37	498
6	6·562	6	3	1,282	70	76·554	70	43	874
7	7·655	7	4	615	80	87·491	80	49	1,251
8	8·749	8	4	1,709	90	98·427	90	55	1,627
9	9·843	9	5	1,043	100	109·363	100	62	243
10	10·936	10	6	376					

SQUARE MEASURE.

The Table of Length can be used thus—

100 sq. millimetres = 1 sq. centimetre	100 sq. metres = 1 sq. decametre
100 sq. centimetres = 1 sq. decimetre	100 sq. decametres = 1 sq. hectometre
100 sq. decimetres = 1 sq. metre	100 sq. hectometres = 1 sq. kilometre

But 100 sq. metres = 1 are (the unit for area), so the following table is used :—

SQUARE MEASURE WITH VALUES IN SQUARE YARDS.

10 centiares = 1 deciare	= 11·96033 sq. yds.
10 deciares = 1 are	= 119·6033 "
10 ares = 1 decare	= 1,196·033 "
10 decares = 1 hectare	= 11,960·33 "

The are is equal to ..
- 1,076·4299 sq. ft.
- 119·6033 sq. yds.
- 3·95383 perches
- 0·24711 sq. chain
- 0·09884 rood
- 0·024711 acre

Proportion to Are.

			Acres.	Roods.	Perch.	Sq. yds	Sq. ft.
Centiare	1/100	10·764299 sq. ft.				1	1·7643
Deciare	1/10	107·64299 sq. ft.				11	8·643
Are	1				3	28	7·68
Decare	10	0·988457 rood			39	16	2·54
Hectare	100	{ 2·471143 acres, or { 2 acres, 2,280·3326 sq. yds. }	2	1	3	11	5·1

40 ares are about an acre.

HECTARES INTO ACRES, ROODS, PERCHES.

Hectares.				Acres.	Roods.	Perches.	Hectares.				Acres.	Roods.	Perches.
1				2	1	35	20				49	1	28
2				4	3	31	30				74	0	21
3				7	1	26	40				98	3	15
4				9	3	22	50				123	2	9
5				12	1	17	60				148	1	3
6				14	3	12	70				172	3	37
7				17	1	8	80				197	2	38
8				19	3	3	90				222	1	24
9				22	0	38	100				247	0	18
10				24	2	34							

LIQUID MEASURE WITH VALUE IN PINTS.

The litre is the unit for liquid measure and = 1 cubic decimetre, so that 1 kilolitre = 1 cubic metre or 1 stere

				Pints.
10 centilitres	=	1 decilitre	=	0·1760773
10 decilitres	=	1 litre	=	1·760773
10 litres	=	1 decalitre	=	17·60773
10 decalitres	=	1 hectolitre	=	176·0773
10 hectolitres	=	1 kilolitre	=	1,760·773

The litre is equal to
- 61·02705 cubic inches
- 1·76077 imperial pint
- 0·22009 imperial gallon
- 0·02751 imperial bushel

Proportion to Litre.

Millimetre	$\frac{1}{1000}$	.. 568 to a pint
Centilitre	$\frac{1}{100}$	.. about 56 ,,

Proportion to Litre.

Decilitre	$\frac{1}{10}$	.. about 5½ to a pint
Litre	1	.. ,, 1⅞ pint

		Liquid.		Dry.					
		Gallons.	Pints.	Qrs.	Bush.	Pecks.	Gals.	Quarts.	Pints.
Decalitre	10	2	1·6077			1	0	0	1·6077
Hectolitre	100	22	0·07744		2	3	0	[illegible]	0·077
Kilolitre	1,000	220	0 7744	3	3	2	0	0	0·77
Myrialitre	10,000	2,200	7·744	34	3	0	0	3	1·7

4·5434 litres = 1 gallon.

Litres into Gallons and Quarts.

Litres.	Gals	Qts.	Litres.	Gals.	Qts.
1	0	0·880	20	4	1·608
2	0	1·761	30	6	2·412
3	0	2·641	40	8	3·215
4	0	3·521	50	11	0·019
5	1	0·402	60	13	0·823
6	1	1·282	70	15	1·627
7	1	2·163	80	17	2·431
8	1	3·041	90	19	3·235
9	1	3·923	100	22	0·039
10	2	0·804			

Hectolitres into Quarters and Bushels.

Hectolitres.	Qts.	Bush.	Hectolitres.	Qts.	Bush.
1	0	2·751	20	6	7·024
2	0	5·502	30	10	2·536
3	1	0·254	40	13	6·048
4	1	3·005	50	17	1·560
5	1	5·756	60	20	5·072
6	2	0·507	70	24	0·585
7	2	3·258	80	27	4·097
8	2	6·018	90	30	7·609
9	3	0·761	100	34	3·121
10	3	3·512			

WEIGHTS.

The unit of weight is the gramme, and is the weight of a cubic centimetre of water at its greatest density, 4°C., or 39°F.

MEASURE OF WEIGHT WITH VALUES IN GRAINS.

10 milligrammes ..	=	1 centigramme ..	=	0·15432348
10 centigrammes ..	=	1 decigramme ..	=	1·5432348
10 decigrammes ..	=	1 gramme	=	15·432348
10 grammes ..	=	1 decagramme ..	=	154·32348
10 decagrammes ..	=	1 hectogramme ..	=	1,543·2348
10 hectogrammes ..	=	1 kilogramme ..	=	15,432·348
10 kilogrammes ..	=	1 myriagramme ..	=	154,323·48
10 myriagrammes ..	=	1 quintal	=	1,543,234·8
10 quintals ..	=	1 millier	=	15,432,348

The gramme weighs .. ..
- 15·4323 grains
- 0·0321507 ounce, troy
- 0·0352739 ounce, avoirdupois
- 0·0026792 pound, troy
- 0·00220462 pound, avoirdupois

	Proportion to Gramme.	Avoirdupois. Cwt.	Qr.	Lb.	Oz.	Dram.	Troy. Lb.	Oz.	Dwt.	Grain.
Milligramme	$\frac{1}{1000}$	..	..	..	..	0·00056438	..	..	..	0·0154
Centigramme	$\frac{1}{100}$	..	..	..	..	0·0056438	..	..	..	0·1543
Decigramme	$\frac{1}{10}$	..	..	..	..	0·056438	..	..	..	1·5432
Gramme	1	..	..	..	..	0·56438	..	..	..	15·4323
Decagramme	10	..	..	..	..	5·6438	..	..	6	10 3234
Hectogramme	100	..	..	..	8	8·4383	..	3	4	7·2347
Kilogramme	1,000	..	..	2	3	3·383	2	8	3	0·347
Myriagramme	10,000	..	..	22	0	11·8304	26	9	10	3·47
Quintal	100,000	1	3	24	7	6·304	267	11	1	10·7
Millier	1,000,000	19	2	20	9	15·04	2,679	2	14	12

A cubic decimetre or a litre weighs 1 kilogramme or 1,000 grammes.
The millier, called tonneau de mer, is the ton of shipping.

KILOGRAMMES INTO CWTS., QRS., LBS., OZ.

Kilogrammes	Cwts.	Qrs	Lbs.	Oz.	Kilogrammes	Cwts.	Qrs.	Lbs.	Oz.
1	0	0	2	3¼	20	0	1	16	1½
2	0	0	4	6½	30	0	2	10	2½
3	0	0	6	9¾	40	0	3	4	3
4	0	0	8	13	50	0	3	26	3¾
5	0	0	11	¼	60	1	0	20	4½
6	0	0	13	3½	70	1	1	14	5¼
7	0	0	15	7	80	1	2	8	6
8	0	0	17	10¼	90	1	3	2	6½
9	0	0	19	13½	100	1	3	24	7
10	0	0	22	¾					

Printed in the USA
CPSIA information can be obtained
at www.ICGtesting.com
LVHW042135161223
766691LV00045B/534